SUSTAINABLE DEVELOPMENT OF ORGANIC AGRICULTURE

Historical Perspectives

SUSTAINABLE DEVELOPMENT OF ORGANIC AGRICULTURE

Historical Perspectives

Edited by

Kim Etingoff

Apple Academic Press Inc.
3333 Mistwell Crescent
Oakville, ON L6L 0A2
Canada

Apple Academic Press Inc.
9 Spinnaker Way
Waretown, NJ 08758
USA

First issued in paperback 2021

Exclusive worldwide distribution by CRC Press, a member of Taylor & Francis Group

ISBN 13: 978-1-77-463690-9 (pbk)
ISBN 13: 978-1-77-188483-9 (hbk)

Library and Archives Canada Cataloguing in Publication

Sustainable development of organic agriculture : historical perspectives / edited by Kim Etingoff.

Includes bibliographical references and index.
Issued in print and electronic formats.

ISBN 978-1-77188-483-9 (hardcover).--ISBN 978-1-77188-484-6 (pdf)

1. Organic farming--Case studies. 2. Sustainable agriculture--Case studies. I. Etingoff, Kim, author, editor

S605.5.S87 2017 631.5'84 C2016-906573-1 C2016-906574-X

Library of Congress Cataloging-in-Publication Data

Names: Etingoff, Kim.
Title: Sustainable development of organic agriculture : historical perspectives / editor, Kim Etingoff.
Description: Waretown, NJ : Apple Academic Press, 2017. | Includes bibliographical references and index.
Identifiers: LCCN 2016050046 (print) | LCCN 2016052332 (ebook) | ISBN 9781771884839 (hardcover : alk. paper) | ISBN 9781315365800 (ebook)
Subjects: LCSH: Organic farming. | Sustainable agriculture.
Classification: LCC S605.5 .S985 2017 (print) | LCC S605.5 (ebook) | DDC 631.5/84--dc23
LC record available at https://lccn.loc.gov/2016050046

Apple Academic Press also publishes its books in a variety of electronic formats. Some content that appears in print may not be available in electronic format. For information about Apple Academic Press products, visit our website at **www.appleacademicpress.com** and the CRC Press website at **www.crcpress.com**

About the Editor

Kim Etingoff

Kim Etingoff has a Tufts University terminal master's degree in Urban and Environmental Policy and Planning. Her recent experience includes researching with Initiative for a Competitive Inner City, a report on food resiliency within the city of Boston. She worked in partnership with the Dudley Street Neighborhood Initiative and Alternatives for Community and Environment to support a community food-planning process based in a Boston neighborhood, which was oriented toward creating a vehicle for community action around urban food issues, providing extensive background research to ground the resident-led planning process. She has worked in the Boston Mayor's Office of New Urban Mechanics, and has also coordinated and developed programs in urban agriculture and nutrition education. In addition, she has many years of experience researching, writing, and editing educational and academic books on environmental and food issues.

Contents

List of Contributors ... *ix*

Acknowledgments and How to Cite ... *xiii*

Introduction ... *xv*

Part I: Past and Current Perspectives ... **1**

1. **The Organic Food Philosophy: A Qualitative Exploration of the Practices, Values, and Beliefs of Dutch Organic Consumers Within a Cultural–Historical Frame** ... **3**

   Hanna Schösler, Joop de Boer, and Jan J. Boersema

2. **Organic Farming: The Arrival and Uptake of the Dissident Agriculture Meme in Australia** ... **31**

   John Paull

3. **Overview of the Global Spread of Conservation Agriculture** ... **53**

   Theodor Friedrich, Rolf Derpsch, and Amir Kassam

4. **The Transition from Green to Evergreen Revolution** ... **69**

   M. S. Swaminathan and P. C. Kesavan

5. **A Review of Long-Term Organic Comparison Trials in the U.S.** ... **79**

   Kathleen Delate, Cynthia Cambardella, Craig Chase, and Robert Turnbull

Part II: Preparing for the Future ... **97**

6. **Keeping the Actors in the Organic System Learning: The Role of Organic Farmers' Experiments** ... **99**

   Christian R. Vogl, Susanne Kummer, Friedrich Leitgeb, Christoph Schunko, and Magdalena Aigner

7. **Supporting Innovation in Organic Agriculture: A European Perspective Using Experience from the SOLID Project** ... **115**

   Susanne Padel, Mette Vaarst, and Konstantinos Zaralis

8. **Organic Farming and Sustainable Agriculture in Malaysia: Organic Farmers' Challenges Towards Adoption** ... **135**

   Neda Tiraieyari, Azimi Hamzah, and Bahaman Abu Samah

9. Are Organic Standards Sufficient to Ensure Sustainable Agriculture? Lessons From New Zealand's ARGOS and Sustainability Dashboard Projects.......... 147

Charles Merfield, Henrik Moller, Jon Manhire, Chris Rosin, Solis Norton, Peter Carey, Lesley Hunt, John Reid, John Fairweather, Jayson Benge, Isabelle Le Quellec, Hugh Campbell, David Lucock, Caroline Saunders, Catriona MacLeod, Andrew Barber, and Alaric McCarthy

10. An Ecologically Sustainable Approach to Agricultural Production Intensification: Global Perspectives and Developments 171

Amir Kassam and Theodor Friedrich

Part III: Annotated Bibliographies for Organic and Sustainable Agriculture 185

11. Tracing the Evolution of Organic/Sustainable Agriculture: A Selected and Annotated Bibliography.......... 187

Mary V. Gold and Jane Potter Gates

12. Twenty-First-Century Organic and Sustainable Farming: A Brief Annotated Bibliography.......... 293

Kim Etingoff

Keywords.......... 299

Author Notes 301

Index.......... 303

List of Contributors

Bahaman Abu Samah
Institute for Social Science Studies, University Putra Malaysia, Malaysia

Magdalena Aigner
Division of Organic Farming, Department of Sustainable Agricultural Systems, University of Natural Resources and Life Sciences Vienna (BOKU), Austria

Andrew Barber
The AgriBusiness Group, PO Box 4354, Christchurch, New Zealand

Jayson Benge
The AgriBusiness Group, PO Box 4354, Christchurch, New Zealand

Cynthia Cambardella
USDA-ARS, National Laboratory for Agriculture & the Environment, Ames, Iowa, USA

Hugh Campbell
Centre for Sustainability, University of Otago, PO Box 56, Dunedin, New Zealand

Peter Carey
Land Research Services, Lincoln, New Zealand

Craig Chase
Leopold Center for Sustainable Agriculture, Ames, Iowa, USA

Kathleen Delate
Department of Agronomy, Iowa State University, Ames, Iowa, USA

R. Derpsch
International Consultant for Conservation Agriculture/No-till,
C.C. 13223 Shopping del Sol, Asunción, Paraguay

Kim Etingoff
Massachusetts State Health Department, Boston, Massachusetts, USA

John Fairweather
The Agribusiness and Economic Research Unit, Lincoln University, New Zealand

T. Friedrich
Food and Agriculture Organization (FAO) of the United Nations,
Viale delle Terme di Caracalla, 00153 Rome, Italy

Jane Potter Gates
Alternative Farming Systems Information Center, National Agricultural Library, U.S. Department of Agriculture, Washington DC, USA

Mary V. Gold
Alternative Farming Systems Information Center, National Agricultural Library, U.S. Department of Agriculture, Washington DC, USA

Azimi Hamzah
Institute for Social Science Studies, University Putra Malaysia, Malaysia

Lesley Hunt
The Agribusiness and Economic Research Unit, Lincoln University, New Zealand

Amir Kassam
School of Agriculture, Policy and Development, University of Reading, Reading RG6 6AR, UK

Susanne Kummer
Division of Organic Farming, Department of Sustainable Agricultural Systems, University of Natural Resources and Life Sciences Vienna (BOKU), Austria

Friedrich Leitgeb
Division of Organic Farming, Department of Sustainable Agricultural Systems, University of Natural Resources and Life Sciences Vienna (BOKU), Austria

Isabelle Le Quellec
The AgriBusiness Group, PO Box 4354, Christchurch, New Zealand

David Lucock
The AgriBusiness Group, PO Box 4354, Christchurch, New Zealand

Catriona MacLeod
Landcare Research, Lincoln, New Zealand

Jon Manhire
The AgriBusiness Group, PO Box 4354, Christchurch, New Zealand

Alaric McCarthy
Centre for Sustainability, University of Otago, PO Box 56, Dunedin, New Zealand

Charles Merfield
BHU Future Farming Centre, PO Box 69113, Lincoln 7646, Canterbury, New Zealand

Henrik Moller
Centre for Sustainability, University of Otago, PO Box 56, Dunedin, New Zealand

Solis Norton
Centre for Sustainability, University of Otago, PO Box 56, Dunedin, New Zealand

Susanne Padel
The Organic Research Centre, Near Newbury, United Kingdom

John Paull
School of Land & Food, University of Tasmania, Australia

John Reid
Ngāi Tahu Research Centre, University of Canterbury, Private Bag 14-800, Christchurch 8140, New Zealand

Chris Rosin
Honorary Fellow, Center for Integrated Agricultural Systems, University of Wisconsin, Madison, 1535 Observatory Dr, Madison, WI 53706, USA

Christoph Schunko
Division of Organic Farming, Department of Sustainable Agricultural Systems, University of Natural Resources and Life Sciences Vienna (BOKU), Austria

M. S. Swaminathan
M. S. Swaminathan Research Foundatio,n Third Cross Street, Taramani Institutional Area, Chennai, 600113, India

Neda Tiraieyari
Institute for Social Science Studies, University Putra Malaysia, Malaysia

Robert Turnbull
Department of Agronomy, Iowa State University, Ames, Iowa, USA

Mette Vaarst
International Centre for Research in Organic Food Systems, Tjele, Denmark

Christian R. Vogl
Division of Organic Farming, Department of Sustainable Agricultural Systems, University of Natural Resources and Life Sciences Vienna (BOKU), Austria

Konstantinos Zaralis
The Organic Research Centre, Near Newbury, United Kingdom

Acknowledgment and How to Cite

The editor and publisher thank each of the authors who contributed to this book. Many of the chapters in this book were previously published elsewhere. To cite the work contained in this book and to view the individual permissions, please refer to the citation at the beginning of each chapter. The editor carefully selected each chapter individually to provide a nuanced look at the development of organic and sustainable agriculture, with the conclusion that organic is not enough to be sustainable.

The chapters included are broken into three sections, which describe the following topics:

Part I provides past and current perspectives on organic and sustainable agriculture.

- Chapter 1 focuses on the cultural dimension of organic consumption, because it may help to explain in more depth what makes alternative food choices so valuable to these consumers.
- Chapter 2 investigates the arrival, reception and uptake in Australia of the organic farming.
- Chapter 3 moves past organic agriculture to sustainable crop production that both reduces the impact of climate change on crop production and also mitigates the factors that cause climate change by reducing emissions and by contributing to carbon sequestration in soils.
- Chapter 4 indicates the differences between the "green revolution" and the "evergreen revolution" that is necessary to fight both famine and rural poverty in an ecofriendly and socially inclusive manner.
- Chapter 5 examines six of the oldest grain-crop-based organic comparison experiments in the United States in order to demonstrate the unique contributions of each site and the usefulness of these sites in communicating agronomic, as well as environmental and economic outcomes from organic agroecosystems, to both producers and policymakers.

Part II focuses on preparing for the future.

- Chapter 6 proposes the concept of farmers' experiments as one option for describing the creative process that might lead to farmers' innovations.
- Chapter 7 explores how innovation occurs within the organic sector in Europe and how this process can be further supported, using framework of innovation systems and experiences from the ongoing project Sustainable Organic Low-Input Dairying (SOLID).
- Chapter 8 highlights the challenges of organic farmers in Malaysia.
- Chapter 9 examines this question: Are organic standards sufficient to secure ecosystem services in the broader way that the United Nations and other frameworks are now promulgating as necessary to ensure sustainability and resilience of farming?
- Chapter 10 proposes that the future for farmers must be conservation agriculture, with special emphasis placed on the need for a mindset change among farmers especially in traditional farming communities. The authors stress the importance of involving all stakeholders to apply a holistic approach in promoting conservation agriculture that is just as much farmer driven as it is science driven.

Part III contains annotated bibliographies for the history and development of organic and sustainable agriculture.

- Chapter 11 traces the development of organic and sustainable agriculture through more than 500 years, ending with the early twenty-first century.
- Chapter 12 includes references for the past decade.

Introduction

Part I of this book delves into the ways that people have approached organic agriculture in sociological, scientific, and economic terms. Part II looks ahead to the future of organic agriculture, presenting opportunities for further progress. Part III consists of an extensive bibliography chronologically developing the progress of organic and sustainable agriculture over two thousand years.

Food consumption has been identified as a realm of key importance for progressing the world towards more sustainable consumption overall. Consumers have the option to choose organic food as a visible product of more ecologically integrated farming methods and, in general, more carefully produced food. Chapter 1 aims to investigate the choice for organic from a cultural–historical perspective and aims to reveal the food philosophy of current organic consumers in The Netherlands. A concise history of the organic food movement is provided going back to the German Lebensreform and the American Natural Foods Movement. The authors discuss themes such as the wish to return to a more natural lifestyle, distancing from materialistic lifestyles, and reverting to a more meaningful moral life. Based on a number of in-depth interviews, the study illustrates that these themes are still of influence among current organic consumers who additionally raised the importance of connectedness to nature, awareness, and purity. The authors argue that their values are shared by a much larger part of Dutch society than those currently shopping for organic food. Strengthening these cultural values in the context of more sustainable food choices may help to expand the amount of organic consumers and hereby aid a transition towards more sustainable consumption.

Just four years elapsed between the coining of the term "organic farming" and the founding of an association devoted to the advocacy of organic farming. The world's first association devoted to the promotion and proliferation of organic agriculture, the Australian Organic Farming and Gardening Society (AOFGS), was founded in Sydney, Australia, in October 1944. It is a geographically surprising sequel to the coining of the term "organic farming" by Lord Northbourne and its first appearance in wartime Britain. Northbourne's manifesto of organic farming, *Look to the Land,* was published in London in May 1940. When the AOFGS published a periodical, the *Organic Farming Digest,* it was the first association to publish an organics advocacy journal. Chapter 2 addresses the question of how the "organic farming" meme arrived in Australia. Candidates for influencing the founders of the AOFGS were (a) Lord Northbourne's 1940 book, and/or (b) perhaps the derivative periodical *Organic Farming and Gardening* published in the USA by Jerome Rodale with its first issue dated May 1942, and (c) perhaps also the earlier book, *Biodynamic Farming and Gardening* by Dr. Ehrenfried Pfeiffer which was published in 1938 in multiple editions (in London, New York, Italy, Switzerland and the Netherlands) which set out to introduce biodynamic agriculture to a broad audience. The archives and records of the AOFGS have not been located, and, in their absence, newspapers of the period 1938 to October 1944 (and through the period of the AOFGS, i.e. October 1944 to January 1955) were searched for references to these three potential sources of influence. Pfeiffer and/or his book received two mentions in the Australian press in the pre-AOFGS period (in 1939 and 1942). Rodale and/or his periodical were not reported in the Australian press in the pre-AOFGS period. Northbourne and/or his book were reported in the Australian press as early as July 1940, and up the founding of the AOFGS, there were 14 Northbourne mentions in the Australian press (all of them favourable or neutral) across four states: South Australia (SA) (n=6); New South Wales (NSW) (n=4); Western Australia (WA) (n=3); and Queensland (QLD) (n=1). The conclusion drawn is that in adopting the term 'organic farming', the AOFGS was informed primarily, and perhaps exclusively, by Northbourne's book *Look to the Land.*

The global empirical evidence shows that farmer-led transformation of agricultural production systems based on Conservation Agriculture (CA) principles is already occurring and gathering momentum worldwide as a new paradigm for the 21st century. The data presented in chapter 3 is mainly based on estimates made by farmer organizations, agro-industry, and well-informed individuals provide an overview of CA adoption and spread by country, as well as the extent of CA adoption by continent. CA systems, comprising minimum mechanical soil disturbance, organic mulch cover, and crop species diversification, in conjunction with other good practices of crop and production management, are now practiced globally on about 125 M ha in all continents and all agricultural ecologies, including in the various temperate environments. While in 1973/74 CA systems covered only 2.8 M ha worldwide, the area had grown in 1999 to 45 M ha, and by 2003 to 72 M ha. In the last 11 years CA systems have expanded at an average rate of more than 7 M ha per year showing the increased interest of farmers and national governments in this alternate production method. Adoption has been intense in North and South America as well as in Australia and New Zealand, and more recently in Asia and Africa where the awareness and adoption of CA is on the increase. Chapter 3 presents the history of adoption and analyses reasons and actual regional trends for adoption to draw conclusions about future promotion of CA.

In the nineteen sixties, dwarf and semi-dwarf varieties of wheat and rice with long panicles and responding favorably to exogenous inputs of inorganic chemical fertilizers registered dramatic increases in productivity (kg/ha.). While based on the yield consideration for the immediate present, it could be regarded as a Green Revolution, it was really an unsustainable exploitative agriculture over long periods of cultivation. The Green Revolution/exploitative agriculture, as expected, started showing signs of yield fatigue since 1990s owing to depletion of biodiversity and freshwater, and degradation of soil health. Further, the Green Revolution which substantially built the food security at the national level did not provide food security at the individual household levels to hundreds of millions of rural women and men. This was because it did not reduce the famine of rural livelihoods and increase the access (i.e. purchasing power) to food. For these reasons, a 'systems approach'–based 'evergreen revolution' was developed. Its design provides for various forms of eco-agriculture to produce food for 'availability' and for harnessing eco-technologies for sustainable management of resources and creation of market-driven on-farm and non-farm ecoenterprises to enhance 'access' to food at the individual household level of hundreds of millions of rural women and men. The evergreen revolution in the resource-poor small and marginal farms integrates diverse agri-horticultural crops, fodder, farm animals and also capture and culture fisheries in the coastal villages. Further, such small and marginal farms could cultivate biofortified crops as to provide agri-horticultural remedies to nutritional maladies, and use dung to generate methane for cooking purposes. Consequently, no methane is emitted into the atmosphere, and when that is used as cooking gas in the rural households, it saves the women from the drudgery of cutting and collecting fuel wood on one hand, and the trees which sequester carbon on the other. All these aspects are discussed briefly in chapter 4.

In chapter 5, long-term organic farming system trials were established across the U.S. to capture baseline agronomic, economic and environmental data related to organic conversion under varying climatic conditions. These sites have proven useful in providing supporting evidence for successful transition from conventional to organic practices. All experiments chosen for this review were transdisciplinary in nature; analyzed comprehensive system components (productivity, soil health, pest status, and economics); and contained all crops within each rotation and cropping system each year to ensure the most robust analysis. In addition to yield comparisons, necessary for determining the viability of organic operations, ecosystem services, such as soil carbon capture, nutrient cycling, pest suppression, and water quality enhancement, were documented for organic systems in these trials. Outcomes from these long-term trials were critical in elucidating factors underlying less than optimal yields in organic systems, which typically involved inadequate weed management and insufficient soil fertility at certain sites. Finally, these experiments serve as valuable demonstrations of the economic viability of organic systems for farmers and policymakers interested in viewing farm-scale organic operations and crop performance.

The creative process that leads to farmers' innovations is rarely studied or described precisely in agricultural sciences. For academic scientists, obvious limitations of farmers' experiments are, for example, precision, reliability, robustness, accuracy, validity or the correct analysis of cause and effect. Nevertheless, chapter 6 proposes that "farmers' experiments" underpin innovations that keep organic farming locally tuned for sustainability and adaptable to changing economic, social and ecological conditions. The authors first researched the structure and role of farmers' experiments by conducting semi-structured interviews of 47 organic farmers in Austria and 72 organic/agroecology farmers in Cuba in 2007 and 2008. Seventy-six more structured interviews explored the topics and methods used by Austrian farmers that were "trying something." Farmers engaged in activities that can be labeled as farmers' experiments because these activities include considerable planning, manipulating variables, monitoring effects and communicating results. In Austria and Cuba, 487 and 370 individual topics, respectively, were mentioned for experimenting by the respondents. These included topics like the introduction of new species or varieties, testing various ways of commercialization or the testing of alternative remedies. Two-thirds (Austria) and one-third (Cuba) of the farmers who experimented had an explicit mental or written plan before starting. In both countries, the majority of the farmers stated that they set up their experiments first on a small scale and expanded them if the outcome of the experiments was satisfactory. Repetitions were done by running experiments in subsequent years and the majority of the farmers monitored the experiments regularly. In both countries, many experiments were not discrete actions but nested in time and space. For further research on learning and innovation in organic farming, the authors propose an explicit appreciation of farmers' experiments, encouraging further in-depth research on the details of the farmers' experimental process and encouraging the inclusion of farmers' experiments in strategies for innovation in organic and nonorganic farming. Strategic research and innovation agendas for organic farming would benefit from including organic farmers as co-researchers in all steps of the research process in order to encourage co-learning between academic scientists and organic farmers.

Organic farming is recognized as one source for innovation helping agriculture to develop sustainably. However, the understanding of innovation in agriculture is characterized by technical optimism, relying mainly on new inputs and technologies originating from research. Chapter 7 uses the alternative framework of innovation systems describing innovation as the outcome of stakeholder interaction and examples from the SOLID (Sustainable Organic Low-Input Dairying) project to discuss the role of farmers, researchers and knowledge exchange for innovation. The authors used a farmer-led participatory approach to identify problems of organic and low-input dairy farming in Europe and develop and evaluate innovative practices. Experience so far shows that improvements of sustainability can be made through better exploitation of knowledge. For example, it is recognized that optimal utilization of good quality forage is vitally important, but farmers showed a lack of confidence in the reliability of forage production both in quantity and quality. The authors conclude that the systems framework improves the understanding of innovation processes in organic agriculture. Farmer-led research is an effective way to bring together the scientific approach with the farmers' practical and context knowledge in finding solutions to problems experienced by farmers and to develop sustainability.

Sustainable agriculture and organic farming are being promoted by the government as a way of eliminating unsustainable agriculture. Despite the benefits that organic farming brings to farmers and environments, its adoption rate is still low among Malaysian farmers. A study of organic farmers in the Cameron Highlands is reported in chapter 8 to reveal the challenges that have been occurred with regard to adoption of the practice. The results indicate that organic farmers face challenges with regard to land tenure, certification processes, hiring foreign workers, marketing, training and extension services and governmental support. Issues and challenges are discussed, and the authors conclude with some recommendations.

Chapter 9 concludes that organic standards need to account for a broader set of criteria in order to retain claims to "sustainability." Measurements of the ecological, economic and social outcomes from over 96 kiwifruit, sheep/beef and dairy farms in New Zealand between 2004 and 2012 by The Agricultural Research Group on Sustainability (ARGOS) project showed some enhanced ecosystem services from organic agriculture that will assist a "land-sharing" approach for sustainable land management. However, the efficiency of provisioning services is reduced in organic systems and this potentially undermines a "land-sparing" strategy to secure food security and ecosystem services. Other aspects of the farm operation that are not considered in the organic standards sometimes had just as much or even a greater effect on ecosystem services than restriction of chemical inputs and synthetic fertilizers. An organic farming version of the New Zealand Sustainability Dashboard will integrate organic standards and wider agricultural best practice into a broad and multidimensional sustainability assessment framework and package of learning tools. There is huge variation in performance of farms within a given farming system. Therefore improving ecosystem services depends as much on locally tuned learning and adjustments of farm practice on individual farms as on uptake of organic or Integrated Management farming system protocols.

Chapter 10 focuses on conservation agriculture. Ultimately, organic and sustainable agriculture are not synonymous. Conservation agriculture (CA) as such is still a new paradigm of farming that is applied on only a small part of the farmland worldwide (11%), with a perspective of exponential growth and the potential to eventually be the most likely farming paradigm for the future of farming. The root cause of agricultural land degradation and decreasing productivity—as seen in terms of loss of soil health—is our low soil-carbon farming paradigm of intensive tillage which disrupts and debilitates many important soil-mediated ecosystem functions. For the most part agricultural soils in tillage-based farming without organic surface residue protection are becoming de-structured and compacted, exposed to increased runoff and erosion, and soil life and biodiversity is deprived of habitat and starved of organic matter, leading to decrease in soil's biological recuperating capacity. CA is a cropping system based on no or minimum mechanical soil disturbance, permanent organic mulch soil cover, and crop diversification. It is an effective solution to stopping agricultural land degradation, for rehabilitation, and for sustainable crop production intensification. CA is now adopted by large and small farmers on some 125 million hectares across all continents and is spreading at an annual rate of about 7 million hectares. Advantages offered by CA to farmers include better livelihood and income, decrease in financial risks, and climate change adaptability and mitigation. For the small manual farmer, CA offers ultimately up to 50% labour saving, less drudgery, stable yields, and improved food security. To the mechanised farmers CA offers lower fuel use and less machinery and maintenance costs, and reduced inputs and cost of production (including labour when CA involves the use of integrated weed management. In pro-poor development programmes, every effort should be made to help producers adopt CA production systems. This is because CA produces more from less, can be adopted and practiced by small-holder poor farmers, builds on the farmer's own natural resource base, does not entirely depend on purchased derived inputs, and is relatively less costly in the early stages of production intensification.

Although many consumers may understand organic production and methods as a recent phenomenon, it has a long and varied history. This compilation presents an in-depth view spanning past values and practices, present understandings, and potential futures, and covering a range of concrete case studies. It posits organic and sustainable agriculture as parallel, sometimes congruent, sometimes very different movements. The organic movement laid the foundation in the past, the future requires agricultural methods that go beyond organic—that are truly sustainable.

—Kim Etingoff

PART 1
Past and Current Perspectives

CHAPTER 1

The Organic Food Philosophy: A Qualitative Exploration of the Practices, Values, and Beliefs of Dutch Organic Consumers Within a Cultural–Historical Frame

Hanna Schösler, Joop De Boer, and
Jan J. Boersema

1.1 INTRODUCTION

Food consumption has been identified as an area of key importance if the world wants to progress towards more sustainable consumption (Carlsson-Kanyama and González 2009; Stehfest et al. 2009). In Western societies, such as The Netherlands, this implies a transition towards less animal-derived proteins (Aiking 2011; Reijnders and Soret 2003) and, in general, more carefully produced food. This transition will not be easy, however, because the relationships between food producers and consumers are bounded by many economic, cultural, and geographic constraints, and all food seems to be embedded in a contested discourse of knowledge claims (Goodman and DuPuis 2002). Moreover, to consumers, changes towards more sustainable food patterns seem only worthwhile when the changes not only enable their pursuit of lifestyles with a lighter environmental burden but are also perceived as rewarding (de Vries and Petersen 2009). Organic food has the potential to meet these demands because it is more sustainable (Badgley et al. 2007; Thøgersen 2010) and because it has become increasingly popular with consumers all over the world (IFOAM 2011). In order to fully utilize this potential, it may be necessary to better understand the cultural context of the choice for organic food, because the cultural changes that will be needed to shift towards a more sustainable society and associated food choices are profound (Aiking 2011). In the past, several marketing studies have been done to identify consumer segments where market share can be increased (Aertsens et al. 2009; Hughner et al. 2007). Within this line of research, however, it tends to be forgotten that the emergence of organic food is associated with reflexive consumption (Goodman and DuPuis 2002) and cultural changes in Western societies (Campbell 2007). Also, and of particular interest, organic consumers seem to refuse a passive role in the food system. Taking an active role may enable them to resolve the alleged contradiction between environmentally responsible behavior and a satisfying life (Brown and Kasser 2005), for example, by understanding themselves as part of a natural order (Taylor 1989). Hence, in the present paper we have chosen to put the cultural dimension of organic consumption central, because it may help to explain more in-depth what makes alternative food choices so valuable to these consumers. As Crompton (2011) argues, particular cultural values motivate people to express concern about a range of environmental and social problems, and such values are associated with action to tackle these problems. Our ultimate objective is therefore to derive insights that can facilitate the much needed transition towards more sustainable consumption patterns (Crompton 2011; Jackson 2005).

The approach on which our work is based can be characterized as an extended case study, which analyses the practices of particular individuals (i.e., the cases) in light of cultural patterns that have developed over several centuries. The purpose of this approach is to understand the case and its theoretical significance (Small 2009; Yin 2003). For theory development, a cross-case analysis involving about ten individuals may provide a good basis. A key theoretical concept in our understanding of the individual is the personal "food philosophy" that he or she might hold. A food philosophy refers to a cluster of practices, values and beliefs that evolves over a long period of time within a particular cultural context and is shared on a collective level. The notion of a food philosophy is inspired by the concept of a worldview (Naugle 2002) or an inescapable framework (Taylor 1989). These concepts refer to the cultural backdrop against which people orientate themselves on questions of what is good, valuable, admirable, and worthwhile (Hedlund-de Witt 2011). This backdrop, however, is largely implicit and unarticulated, and people may be unaware of its influence or even resist it (Taylor 1989). However, through the interpretation of empirical interview data, one can uncover an underlying coherence or sense that can generate a better understanding of important dimensions of human life (Taylor 1971). Based on the literature, we will explore the food philosophy of the organic movement by providing a concise historical and cultural background. To analyze the personal food philosophies of current organic consumers, we will present findings from qualitative interviews conducted with individuals in The Netherlands.

The interviews were conducted in 2010 as part of a bigger project that investigated food practices, values, and beliefs among the Dutch population. The case selection was based on the sampling for range approach, in which the researcher identifies subcategories within the study's population and interviews a given number of people in each subcategory (Weiss 1994). The subcategories were delineated according to the different food-related orientations identified in previous survey research (de Boer et al. 2007). The representative survey among Dutch citizens indicated four distinct value-orientations towards food based on the degree of involvement with food and a motivational focus on promotion versus prevention.[1] For our research, we focused on the "reflective" orientation, which entails a careful and mindful use of food and a preference for responsible products (i.e. high involvement combined with a prevention focus). The organic consumers were selected to represent this orientation towards food. According to the survey results, the "reflective" orientation can be found among roughly 14% of the Dutch population (de Boer et al. 2009). As mentioned before, however, the present study does not search for statistical significance, but for theoretical significance. A more qualitative, interpretive approach is needed in order to reveal greater depth and meaning of consumer practices (Hughner et al. 2007). By combining insights into the cultural dimensions of the organic movement with insights into the individual's motivation for using organic products, we will try to facilitate a more complete understanding of these consumers' practices, values and beliefs and the potential influence thereof on a more sustainable diet—more precisely, a diet less reliant on meat.

In sum, the present paper is organized into three sections. First, it provides a concise cultural background of organic food. Next, it presents the results of a qualitative interview study with consumers of organic food in The Netherlands.

Finally, the paper discusses the overall relevance of our findings in the context of the transition towards a more sustainable food system.

However, before moving on, we briefly reflect upon the assumption made above that organic food consumption is indeed part of a more sustainable diet that benefits social and environmental systems as well as human health (Lang and Heasman 2004). Organic farming is defined as a holistic production management, which promotes and enhances agro-ecosystem health and avoids the use of synthetic materials to fulfill any specific function within the system (Codex Alimentarius 1999). Furthermore it adheres to the principle of health as a state of holism, self-regulation, regeneration, and balance and is exemplified by Lady Eve Balfour's quote "healthy soil, healthy plants, healthy people" (IFOAM 2011). Despite these desirable goals, there has been some controversy about the degree to which organic production can contribute to sustainability, given the increased amount of land that organic production requires at the cost of nature reserves (Tilman et al. 2001) and the arguable benefits in terms of biodiversity (Hodgson et al. 2010). On the other hand, it has also been shown that organic agriculture is capable of feeding the world sustainably (Badgley et al. 2007), especially if farming practices that mitigate climate change are also sufficiently employed (Badgley and Perfecto 2007; Scherr and Sthapit 2009). In considering the market of organic food, however, fundamental contradictions have been identified between mainstream agro-industrial and alternative movement conventions, because increases in scale and standardization lead to the bifurcation of the organic sector and the watering down of its original values (Buck et al. 1997; Constance et al. 2008). Thus, a globalizing organic agro-food sector risks susceptibility to similar ills it aimed to cure in the first place (Raynolds 2004). For instance, although it is debatable whether organic agriculture as a whole is becoming conventional, there is a growing influence of conventional agro-food commodity chains in certain sub-sectors of organic agriculture in The Netherlands (de Wit and Verhoog 2007). While acknowledging that trends towards conventionalization in the sector and the development of an organic industry add another set of problems to the sustainability debate, it is crucial to also note the sector's role in society. At the very least the trend towards organic can be seen as a valuable driving force that stimulates conventional agriculture to adopt more ecologically integrated methods and inspires consumers to adopt new values and ideals that can give direction to more sustainable food practices (Lang and Heasman 2004). The consumption of organic food has also been discussed in the context of a shift in worldviews that is taking place in the West (Campbell 2007). According to Campbell, healthy and environmentally friendly food consumption is something that is perceived as deeply satisfying and meaningful. The choice for organic food may have an underestimated religious undertone, providing people with purpose in life and a means to reconnect with nature (Campbell 2007; Pilgrim and Pretty 2010). Such orientations towards food may be understood in the context of contemporary spirituality and can have an important role in facilitating the transition towards a more sustainable society (Hedlund-de Witt 2011). It is the food philosophy associated with the trend towards organic that we are particularly interested in.

1.2 A CONCISE BACKGROUND ON THE ORGANIC MOVEMENT

To gain insights into the cultural dimensions of the organic movement, a two-step research approach was used. In the first step, we identify long-term trends in Western culture that have shaped the origins and the development of the organic and natural foods movement. It should be emphasized, however, that it is not possible within the scope of this paper to give a complete and historically accurate description of these topics. In this exploration, we had to limit ourselves to highlighting some major themes. Food has always been an important symbol that can reveal what conceptions of nature our culture affords and how people might derive identity from it (Douglas 1966; Fischler 1988; Montanari 2006). Specifically, we used philosopher Charles Taylor's acclaimed Sources of the self: The making of the modern identity (1989), because it provides a solid background on the development of Western culture and addresses in depth how changing patterns of thinking affected our conceptions of nature and the natural. The second step of our approach summarizes the history of organic food since the nineteenth century. In bringing together these particular works on culture and food, we tried to identify those cultural ideas that distinguish organic consumers and characterize their lifestyles.

In *Sources of the Self. The Making of the Modern Identity*, Taylor suggests that modern Western culture, even though it is now characterized as fragmented and pluralistic, builds essentially on two divergent cultural orientations. These orientations are highly relevant to understanding alternative food philosophies. They can be loosely tied to the period of Enlightenment in Western history and the Romantic era. The Enlightenment inspired patterns of thought that emphasize a rational understanding and scientific reasoning about reality, which implies abstraction and objectification of the world and the natural phenomena one can observe. In this sense, it broke away from the mystical understanding of nature that used to be dominant prior to the Enlightenment (Glacken 1967). Nature, which includes the human body, is understood by constructing a correct representation of it in one's mind, thus making it a neutral object to observe and study (Taylor 1989). This implies that nature has no meaning beyond its function and a value that is only dependent on utility. Reason empowers people to study and observe the natural world, which in turn can lead to ideas that nature can be controlled and manipulated. Taylor (1989) argues that this objectification and instrumentalization of nature leads to a separation from nature and our moral independence from it.

Partly as a reaction to the instrumentalization of nature, Taylor (1989) argues that Western culture turned to creativity, intuition, and expressivity as a means to re-unify with nature. Mankind is then seen as an integral part of a larger order of living things that nourishes human life and creates bonds of solidarity within a mutually sustaining web of life. Even though Romantic religions of nature have died away, Taylor (1989: 384) argues that "the idea of being open to nature within us and without is still a very powerful one that is grounded in the understanding that mankind is part of a larger order of living beings, in the sense that their life springs from there and is sustained from there." This more embodied orientation towards nature inspires thinking that people should be careful and try to do no harm to nature. In this perspective, or even "way of being," nature is included in people's representations of self (Schultz et al. 2004), and forms a part of their identities. As a result, modern culture is characterized by the tension between the two big constellations of ideas (i.e., Enlightenment and Romantic views), and this tension becomes particularly manifest in the form of controversies about sustainability issues (Taylor 1989).

The divergent cultural ideas are also directly reflected in the history of organic food. While nowadays there is a more prominent link between the organic movement and environmental activism forming in the late 1960s (Foss and Larkin 1976), the roots of the organic movement actually run deeper. Ideas around organic farming developed almost independently in German and English speaking countries about a century ago. In Germany it was part of an influential movement that became known as the Lebensreform and consisted of various Reform movements resisting increasing industrialization, use of technology, materialism, and urbanization that were shaping a new way of life. The Reform movement promoted the return to a more natural way of living that consisted of vegetarian diets, physical training, natural medicine and going back to the land (Vogt 2007). Food was important due to its direct link with the natural environment, the agriculture's dramatic mechanization and industrialization, the loss of rural lifestyles and the associated self-sufficiency and independence (Vogt 2007). Countries undergoing similar changes in the food system, such as Germany, The Netherlands, England and the United States, all exhibited cultural responses similar to the Lebensreform (van Otterloo 1983). The Dutch Reform movement was directly triggered by the developments taking place in Germany (ibid).

Many people perceived the dramatic societal changes and the loss of traditional rural lifestyles as a threat to their moral independence. A self-regimented way of living and control over one's body were symbolic in averting this danger (Barlösius 1997). The Reform movement was therefore associated with a moderate, sometimes ascetic lifestyle. It enabled the individual to feel self-determined and to live according to one's own moral and ethical principles, independent from behavioral prescriptions of government and industry. For example, the use of processed food products was avoided on these grounds. Due to its visibility and daily practice, food consumption was an exceptionally suitable domain for individuals to express their commitment to an ethical and self-determined lifestyle purely founded on one's ideals (ibid).

Vegetarianism was an important part of the lifestyle promoted by the Reform movement (Barlösius 1997), as the consumption of meat is traditionally a morally contested practice (Fiddes 1991). Followers also turned against products of the upcoming food industry and banned instant or canned products. Also, natural stimulants, such as tobacco, coffee, alcohol, sugar, and strong spices, were rejected. Instead, raw vegetables and whole-wheat products were preferred (Barlösius 1997). The essential question raised by the reformers was how human needs in general should be satisfied, which explains why the movement was equally concerned, for example, with housing, clothing, and sexuality. The human body was conceived as a nexus of the individual's needs and the constraints of the societal system.

Gusfield (1992) describes cultural changes very similar to the Lebensreform associated with the American "Natural Foods Movement." The movement had its origins in the 1830s, a period of intense religious reawakening and deep concern over the immorality and crime associated with increasing urbanization and the loss of traditional bodies of authority. One of the key reform thinkers was Sylvester Graham, a Presbyterian minister. He opposed modern food technology and considered the unrefined, the coarse, the pure, and the raw to be healthy qualities while the refined, the smooth, the processed, and the cooked, respectively, were objectionable (ibid). He dismissed refined white bread, the icon of the upcoming food industry, because it had less fiber than the common whole-grain breads and was baked outside the home. Stimulating foods, such as meat, coffee, sugar, or alcohol, were equally abject because they were believed to excite the body in an unhealthy manner, just like sexual desires would. What characterized the philosophy was the capital importance of self-discipline and self-control against the temptations surrounding the individual.

The various organic and natural foods movements were not very successful until the 1960s. It was the publication of *Silent Spring* by Rachel Carson (1962) that became a turning point for both the modern organic and environmental movement (Kristiansen and Merfield 2006). *Silent Spring* brought a whole new set of arguments against industrial farming, in addition to those that the organic movement had been pushing for many decades. Several new movements took up the moral stance towards food and continue to promote a more vegetarian diet and consumption of organic food. Hamilton et al. (1995) suggest that this food "alternativism" is often associated with New Age philosophy and a spiritual worldview (see also: Hedlund-de Witt 2011). In fact, various studies claim that natural and health foods can be viewed in a spiritual context (Campbell 2007; van Otterloo 1983, 1999) and can be linked to spiritual practice, such as mindfulness meditation (Jacob et al. 2009). From the very beginning, spirituality was also incorporated in the Lebensreform by reformers like Rudolf Steiner, who laid the spiritual foundations of organic farming (Kirchmann et al. 2008). The steady popularity of his esoteric philosophy, Anthroposophy, illustrates that the movement is still influential today. Nevertheless, people can also identify with "eating green" for more secular reasons (Jamison 2003). The same is true for the feeling of connectedness to nature (Hyland et al. 2010).

In summary, the insights provided above highlight a number of themes that may explain why organic consumption has been characterized as part of a distinctive way of life (Schifferstein and Oude Kamphuis 1998). These themes include a strong resistance towards food industry and technology, because they were perceived to impose consumption patterns that conflict with particular moral norms. Instead, people tried to conserve their independence and self-determination by orienting towards nature within as a source of morality. The inward orientation of their philosophy often led to spiritual associations and a belief that human needs are not only satisfied by material needs. Self-determination was associated with the practice of a moderate lifestyle—the (partial) abstinence from meat and other "unnatural" foods. In the following section, we compare these insights with the food philosophies of contemporary organic consumers by discussing a cluster of themes that emerged from the interviews we conducted with consumers in The Netherlands.

1.3 FOOD PHILOSOPHIES OF CURRENT ORGANIC CONSUMERS

In this section we move on to the findings from the interviews. Using the sampling for range approach (Weiss 1994), we contacted 33 people via different avenues. Thirteen of them were assigned to the subcategory of organic consumers. Organic consumption in The Netherlands is growing steadily, but is still rather low, compared with other European countries (Bakker 2011). The total market share in The Netherlands in 2009 was 2.3% (ibid). Roughly one-third of organic food is sold at specialized organic stores. [2] As we were interested in consumers who are relatively highly involved with food, we secured interviews with ten people we approached in organic stores in two Dutch cities, Amsterdam and Groningen, the former a more metropolitan, big city and the latter a more rural, small city. Other subcategories of participants were acquired from a hobby-cooking club, the Slow Food organization and at regular supermarkets. The data used in this paper come from the participants approached in the organic shops; other participants that mentioned they regularly use organic food were also included. Altogether 9 women and 4 men participated, varying in age from 18 to 76. Given that women shop more often in organic shops (Hughner et al. 2007), this distribution is acceptable; even so, women were also somewhat more willing to participate. The participants' level of education was relatively high (ten had graduated from university), but they were not particularly wealthy. About half of them were self-employed and had an artistic or creative background. It seemed that their daily routines were comparatively flexible, enabling them to visit farmers' markets during the day or prepare a midday meal at home.

The interviews were introduced as a study on food practices in The Netherlands. There was no prior mentioning of themes relevant to the objectives of this study, such as environmental sustainability or organic food consumption. The researchers engaged participants in conversations aided by some simple questions asking them to describe what they had eaten the day prior to the interview, how they had prepared their meals, and how they shopped for food. These questions were only meant to start the conversation, and participants were allowed to develop their own stories from there, introducing topics that were relevant to them. The interviewers limited their interference to posing questions, inviting participants to further engage in topics that they had brought up. The conversations lasted roughly an hour and were held, if possible, in participants' homes or, otherwise, in a public space. They were taped and transcribed verbatim. The real names of respondents are not provided to ensure their anonymity.

The interviews were analyzed according to the grounded theory approach (Charmaz 2006). This approach encourages the researcher to learn what participants' lives are like and to be sensitive to how they explain their statements and actions. Subsequently, she constructs a theory that is "grounded" in the data, instead of using preconceived, logically deduced hypotheses (Glaser and Strauss 2009). The analytical process involved coding the interview material and constructing conceptual categories from the emerging codes. The analysis of the interview data gave rise to three analytical themes that shed light on the food philosophies of organic consumers in The Netherlands. First, we discuss participants' feeling of connectedness with nature. Second, we discuss the notion of awareness. Third, we explain the value of purity.

1.3.1 CONNECTEDNESS WITH NATURE

Participants expressed a philosophy of "doing what feels natural." Their concern for the naturalness of food made the choice of organic and seasonal foods attractive. They described feeling connected with nature, which triggered feelings of care and responsibility for animals and the natural environment. Nature, however, was not perceived as a separate entity. Rather, participants felt an integral part of nature. Care for nature, therefore, also meant to care for one's physical and mental health, as well as striving for vitality and overall well-being. For example, participants expressed their sense of connection by expressing how season changes and other natural processes correspond to changes in their physical and mental constitution, such as the following participant.

> At the farmers' market, there's a clear offer of the season [...] I find it interesting to do something with the cabbage the moment it's there in wintertime, because I find it fits with the moment, because I have different needs and, then, I like to eat differently. (Mary)

Through their connectedness with nature, participants explained their discovery of the various interdependencies of food and nature. They also became aware of the farmers that farm their vegetables, and they became more sensitive to the issues of familiarity, trust and geographical vicinity.

> Vegetables I buy organic. I have a veggie box. [...] To me it's important that it's farmed with care and that it has travelled as little as possible. And that it's as seasonal as possible, that it comes from a familiar environment [...] I try to think about the consequences of my consumption for the rest, for the environment [...] I think, first came the environment and gradually I've created a connection with the farmers, because it's nice that he knows about us, and you see him every week, and now there is a strong social tie. (John)

One participant, who was also a practitioner of Japanese yoga, explained that it is part of her food philosophy to eat food that is native to her home region and seasonal. She explained that the natural environment influences her inner constitution, and the consumption of food is a vital mediator in this. As various foods have different effects on the body, her goal is always to achieve a balanced constitution by matching the food she selects to the needs of her body. The participant described how she uses her feeling and intuition to access this source of knowledge.

> Your constitution is also partly a result of the weather or the water that you drink. The vegetables that grow here in wintertime, like root celery, are typically warming vegetables. So, that's perfect, because that's exactly what we need then. So, it's natural to eat what's in season here and now [...] I grow physically and mentally stronger, simply because I eat the food that's compatible with my momentary constitution. [...] I eat based on my perception of my own body [...] It is very intuitive, actually. (Katie)

Another participant, who felt inspired by ideas from the macrobiotic and Chinese food philosophies, described a similar connection and a longing for a more intuitive relationship with nature. She argued that people have lost part of their connectedness to nature and, thus, also their intuition about what is the right way to eat.

> The philosophy is that you're one with the cosmos, with the environment. So your food should be seasonal [...] and you try to eat the food that belongs with the climate you live in. I'm not so strict.

When asked why this philosophy appeals to her, she replied:

I like its intangible character. Centuries back, humans had to live with nature; they were dependent on it and adapted to it completely: with the seasons, with the moon. And all this knowledge has been lost. [...] In China it's still more alive, but Europeans also had it. [...] It's a certain feeling about how things need to be done that you cannot explain. But, in our society, this feeling with nature and your environment has weakened. I find it really interesting to try and get [this feeling] back. (Sally)

The connectedness with nature was also evident in people's concern for animals. All participants watched their meat consumption closely and had considerable concern about animal welfare and the inhumane treatment of animals in the agricultural industry. Most of them were or had for periods in their lives been vegetarians. All reported cooking vegetarian food regularly, as well as frequently buying organic meat–meat that is produced in a more responsible manner. If they find buying organic meat too expensive or if it is unavailable, they prepare vegetarian meals. Participants felt that they should eat meat in moderation, and they often doubted the healthiness of regular meat consumption. On top of this, animals were generally seen as sentient fellow creatures with a right to live under natural circumstances, such as those organic farms try to provide. Therefore, the consumption of organic meat was an acceptable alternative for participants. They strongly opposed intensive livestock farming systems, because, to them, the animals are treated like a commodity.

I don't like the fact that animals are seen as products. Maybe that's not the worst...but I think you have to treat animals differently from a bag of cookies. It's hard to explain, but it just feels wrong to me. (Lauren)

We do eat meat, but not regularly. And if we do, it's always organic. [...] What I find really important is the care for the animal. (Mary)

1.3.2 AWARENESS

Cooking and eating, especially with family and friends, were often described as a crucial moment of tranquillity and awareness in a busy life. Participants associated the moments that they can engage with food with a sense of well-being and happiness. They described their enjoyment in focusing on activities such as the food preparation, setting the table, making the plates look attractive, and eating the food. To them, these moments are in contrast to other daily activities, in which they often feel rushed and superficially engaged. Participants described a heightened awareness of their surroundings, as well as an awareness of their feelings and emotions.

> I feel happy when I cook, when I have the time to do that [...] Enjoying is not only related to food, though, it's more about what happens here around me. In the evening I have the sun here and then, in combination with being outside, the tranquillity. Sitting here at the table and simply eating something tasty, that's what makes me happy. [...] To me, that's the ultimate pleasure: to find the peace and time to have awareness for that. (Mary)

> It has a lot to do with attention and love. [...] I try to really make contact with food. (Emmy)

> It can be really nice to enjoy food together, but I can also do it alone: when I'm really in the moment and enjoy what I eat or what I do in that moment, without really thinking about it. I mean, my head's always occupied, so I really enjoy when there are no thoughts, when I'm fully engaged in the moment. Of course, it should be a pleasant moment. Yes, that gives me peace and relaxation. (Lauren)

As these participants describe, their moments of awareness and attention to food were often qualified by the absence of thought and a feeling of being immersed in the activity of cooking or eating. This engagement and intense experience of the moment was something that fulfilled them with joy and peace. As the participant describes in the second quote, she establishes a sense of connection with her food by giving it attention.

Participants' awareness of the present moment made the entire context of a meal more salient. They experienced the sensual qualities of a meal: what the food looks like, what it smells and feels like, and what are the particular circumstances of the moment. All these factors contributed to the satisfaction they could derive from a meal. Likewise, they reported that their enjoyment of food was hampered when there was no time to pay attention.

> What I really hate is rush. If there's rush, then all enjoyment is gone. That's really important. Then you don't see things anymore and you don't taste them anymore. (Thomas)

> Engaging with food is the ultimate enjoyment for me, to find the tranquillity to have awareness for it. And stress or unhappiness I associate with having to eat an instant pizza, when I have no other choice. (Mary)

A heightened sensitivity to how one's body responds to food was also a dimension of participants' awareness. They frequently stated they rely on their senses to tell them what food they should eat. They listened to their bodies, when they wanted to find out how they should eat to feel good.

> For a few years I didn't eat meat. I didn't react well to it, so I changed my food pattern. I felt better [...] I noticed that my body responds in a certain way to everything I eat. So, if you eat something and it gives you stomach ache, you don't want to eat that anymore. (Mary)

> You adjust what you eat to your constitution [...] you judge your constitution by sensing what food does to you. So, if you take the energy from food, if you feel that something warms you up, you get a warm tummy, or often I notice my hands getting warm. (Katie)

This heightened awareness also included sensitivity towards one's emotional responses. Participants described how particular food-related experiences—either pleasant or disturbing—made such a profound impression on them that they had a sustained influence on their food practices.

> I visited a slaughterhouse a few times when I was 16. I have two uncles who are butchers. I saw how the cows got a pin shot in the head and the pigs were electrocuted so quickly and immediately hung on the hook while all is still moving. That gave such an impact that shortly after I stopped eating meat. (Thomas)

> I used to have a Scottish boyfriend. He made a lot of things himself, baked his own bread, all very idyllic. He was a fisherman by profession and through him I saw and learned about the fish, about the sea, about the salmon, the fish farms and the consequences [...] and I guess because of him, for me now the only alternative is to choose organic meat and sustainable fish. It's got to do with being engaged with your personal environment, what happens around you. (Mary)

> There seemed to be a link between awareness and the intensity of memories that people described in relation with food. All participants had vibrant memories of formative experiences related to food, which they described in colorful details. Obviously, food consumption was often intimately tied to their emotional experiences and, therefore, left a deep mark on their memories.

1.3.3 PURITY

Participants had developed particular strategies to decide how to eat and what is good to eat. Central to these strategies was the participants' self-determination and the idea that they behave according to their personal values and their individual intuition. They tended to have the opinion that "we" don't know what we are eating because food producers mix substances together and thereby obscure people's choices. Since it has become extremely difficult for consumers to judge the quality and composition of the food they buy, participants categorized food according to their understanding of purity (and related concepts, such as simple, basic, whole, and raw). In what follows, we scrutinize in more detail what participants meant by these qualifications and how they enacted them in their practices. Purity was associated with food in a material sense, but also in an immaterial sense, as it referred to the moral purity of a particular food choice.

> I like my food to be pure. I cook with few spices, so that the original taste of the product is preserved. The product remains itself, and you can really taste it. (Thomas)

> I used to put too many things together, and then you don't taste the pure flavour. So, I went back to cooking pure food. I never buy instant stuff. If I prepare a sauce, I simply start from scratch– that's more pure [...] As soon as I lose myself in all kinds of ingredients that I don't understand, the more processed things are, the further estranged from the original product, the less attractive I find them to be. (Mary)

Participants associated purity with making sure that the essence of the food is preserved. Thus, authenticity and originality were important, in terms of sensual food qualities like taste, appearance, smell, and feel. To preserve this essence, participants kept meals simple and ingredients few. Excessive use of spices, for example, was believed to obscure the true identity of the food. As the second citation illustrates, preparation from scratch was also important, as it helped participants to be aware of all ingredients. Therefore, when shopping, participants searched for raw foods, and they avoided processed foods, which were associated with artificial preservatives, chemical residues, E-numbers and added sugars. Also, the number of ingredients in a product served as an indicator of its purity.

> For example I don't like instant yoghurts with readily added fruit and whatever ingredients there might be. I simply buy plain yoghurt, and then I add whatever I want to add. So, I know what I add. Pure... I prefer to buy the basics and then I'm in charge of mixing things. (Mary)

> Another participant described the difference between the food that simply fills up the body and the food that really has the ability to nourish. He preferred foods that he considered whole and complete in terms of nourishment.
>
> For instance, I hardly go for Chinese take-away. You're stuffed with feed rather than food. You get lots of rice, a tiny bit of vegetables and proportionally lots of meat.

When asked what distinguishes "food," he replied:

> Well, food is the things that take some time to eat and digest, so whole wheat products, rye bread, vegetables, meat, not the things that disappear quickly. (Peter)

To him, the quality of food was expressed in the amount of time that was needed to eat something, as well as the length of time that he felt satisfied afterwards.

Another strategy to preserve the purity of a food is to try and preserve its natural appearance and form.

> I like to serve all ingredients of the meal separately, so that they are visible [...] you see what you eat, nothing is hidden [...] no ornaments or additions that have nothing to do with the original product. (Thomas)
>
> Also, when I have visitors, I don't serve everything hustled in a big pan, but I put things separate. So you can take what you like. Straightforward, elementary, and the food recognizable. I prefer that nothing is hidden! (Helena)
>
> When I say fresh, I don't just mean the due date, but also that it's not in cans. I want to pick the food myself; I have to see it for myself. [...] I think it's really important to touch the food [...] that's why I don't buy canned food; you just can't see it properly. (Sarah)

Participants contrasted "pure," "fresh," "simple," "basic," "plain," "original," or "organic" foods with "estranged," "processed," "instant," "complex," or "canned" foods. One participant explained that these categories of food reflect not only material qualities of food, but are also associated with her moral beliefs regarding what is a good way to live:

> When I talk about unsprayed and organic, I mean something more archetypal, more natural. I feel that we are pushed into more and more artificial circumstances in our society. We're on the wrong track. I think these values that I talked about, just now, awareness, understanding what you need. Of course, one person

can need something more than another person, even with food. I mean, some people have the need to travel around the world and then they should do that. But let's be honest, many people don't have that 'need' they only do it because everybody's doing it. I like when people really work out for themselves, thinking independently, what they actually need. (Emmy)

Thus, the immaterial quality of purity was associated with living a reflective life in which one would try to be modest and sensitive to one's own needs. This idea was also expressed repeatedly in the importance that participants attributed to temperance. Temperance was perceived as a means to be self-determined and to make choices according to one's personal values. For example, participants wanted to express gratitude and respect for food, especially when consuming foods originating from animals.

People have so many desires they want to satisfy immediately, but tasting is important. I mean, a fish has also been an animal; you don't just wolf it down. You have to have some respect for it. That's the kind of temperance we search for. [...] My daughter has a different attitude. [...] She wants instant satisfaction. And if she's hungry, something needs to be done about it, immediately. She's not engaged with taste. (Thomas)

Underlying this temperance was the wish to transcend the bodily desire to eat and to appreciate food on an immaterial level. Part of the enjoyment of food was, therefore, contemplating the meal more fully. This partial shift from the material to the immaterial dimension of food was also represented in the shift of attention from quantity to quality.

To eat organic meat reflects also my conscious choice to consume less, but better quality. It's expensive and that's why I don't eat meat two days a week. That's all connected. So I think you can't view it separately [...] and at the end of the week, I've spent the same amount as I would otherwise. (Michael)

Participants related temperance in their personal food consumption to the boundlessness and overconsumption that they perceived to be the current cultural norm, from which they wanted to distance themselves. The practice of temperance represented to them a shift away from desires and wants and towards their basic needs.

It feels best to me to use just what I need. All this excess and overkill that is the norm now doesn't appeal to me [...] What we often do, when we have leftovers,

> we eat it one day, skip one and then eat it again [...] I find it a sign of no respect to throw out food. (Emmy)

> I think we [Dutch society] have an enormous overconsumption. We use much more than we actually need. (Lauren)

> I buy what I need and try not to be manipulated by all the advertisements and special offers [...] this is what I see many people do: 'it's on sale, so I buy it.' But then at the end of the week, things are past the expiry date and are thrown out. That's a shame. (Peter)

> This massive animal industry, I think it's appalling [...] Raising production is an end in itself! People have to buy different clothes every year, because of fashion, because of the economy. It's insane. (Helena)

These quotes illustrate that participants' particular food choices were associated with a rather critical view of society. They objected to the orientation towards consumerism, and they wanted to resist the manipulative influence of advertising and fashion. The practice of temperance was associated with an orientation inward, towards one's personal needs, that helped participants maintain an intuitive balance. They referred to the importance of being aware of one's body in order to assess the boundaries of what is enough and what is good to eat. This is also tied to the awareness that we described in the previous section.

> I believe you have to eat moderately and healthy [...] I think when you are moderate then you don't fluctuate in weight and in how you feel physically. There is a kind of stability in it. (Theresa)

1.4 THE ORGANIC FOOD PHILOSOPHY AND ITS RELEVANCE TO SUSTAINABILITY

- Based on the history of the organic movement and personal stories of current organic consumers, we can highlight some key elements of the organic food philosophy. A central element of the stories was an intuitively felt connectedness with nature that goes beyond their care for plants and trees. It concerns a reflexive relationship with one's inner nature that is not separate from the "outer" environment and could therefore be described as transcendent (Hyland et al. 2010). In accordance with our findings, Hyland et al. (2010) point out that people usually experience this sense of connection in an all-encompassing way: with regards to nature, places, other people and even the entire universe. This reminds us of Taylor's (1989) description of the wish to re-unify with nature and to feel an integral part of a larger order of living things. As the participants described it, experiencing this special connection requires a subtler language of feeling and awareness, which Taylor (1989) refers to as people's powers of expressivity and creative imagination.

- Tuning into a special connection with inner/outer nature provides people with purpose in life and a means to reconnect with nature (Pilgrim and Pretty 2010). In terms of Taylor (1989), therefore, the organic philosophy fits in with the Romantic worldview. Campbell (2007) has argued that the popularity of this view is connected with an important shift in the Western worldview, where the belief in a distant, personalized god is slowly being replaced by a belief in an undefined immanent divine force that unites humankind, nature, and the cosmos as one. As a consequence, nature becomes sacred and animals are regarded with reverence, while human superiority and dominion of animal life are discarded (Campbell 2007; Verdonk 2009). Naturally, this shift has profound consequences regarding people's views of food, because their food practices are imbued with meaning and the moral dimension of food choice becomes more salient (Campbell 2007).

This interpretation seems to fit with the Dutch context. The Netherlands has been characterized as one of the most secularized countries in Europe (Knippenberg 1998), but at the same time, strong trends towards contemporary spirituality and religious seekership outside the traditional church have been observed (van Otterloo 1999; Versteeg 2007). Food consumption plays an important role in these trends to maintain a healthy body and mind and to improve oneself spiritually (van Otterloo 1999). This orientation may also explain why participants are not very oriented towards asceticism that played an important role in the Lebensreform and the American Natural Foods Movement but is not mentioned in the literature on contemporary spirituality (Hedlund-de Witt 2011). Nevertheless, while some participants emphasized the religious undertone of their practices, more secular interpretations are also possible and may give rise to the same practices (Hyland et al. 2010). Participants could, for example, equally emphasize the importance of care for animals and nature and the solidarity they feel with other people. In general, it seems that a more value-laden approach to food is in line with the times.

Another key element is that the participants shared their self-determined, moral outlook on life. As Hamilton et al. (1995) put it, food practices of people with this orientation are pervaded by "a concern which goes beyond the material, a desire for a meaningful life, a moral life, one which is in harmony and balance, a desire for mental peace, even perhaps simply contentment and happiness." Gusfield (1992) and van Otterloo (1983) add that this orientation can be understood in the context of the individual that wants to protect her (moral) values against the pressures of civilization. A healthy, natural lifestyle and the discipline to abstain from desires that are constantly aroused by a consumption-oriented environment are experienced as part of the good life. Within this context, the relevance of moral themes, such as purity and temperance with regards to food (Kass 1994; Rozin et al. 1997), is evident, and it also emphasizes the timeliness of the ideas associated with the Reform movements.

A limitation of our study is that our description of organic and natural foods movements in Western countries was supplemented with an analysis of the food philosophy of organic consumers from only one of these countries, The Netherlands. This does not enable us to shed more light on the food philosophy of consumers in other Western countries that show similar, but not identical trends of changes in the food system, such as the United States, England and Germany. Although we expect the same basic tension between Enlightenment and Romantic views in these countries, there are many contextual variables that could be important to organic consumers. In particular, differences in transparency between organic and conventional agriculture can be reinforced by contextual factors, such as marketing strategies. In the United States, for instance, organic is framed as a "marketing label," and there seems to be more polarization between the organic and the conventional food chain than in Western Europe (Klintman and Boström 2004). This means that organic consumers may have divergent opinions on the distinctive advantage of organic foods, dependent on the type of market or the maturity of the market in their country (Wier et al. 2008). Future work should examine whether such differences in opinion are also associated with basic differences in food philosophy.

The question now is what is the relevance of the organic food philosophy for a transition towards a more sustainable food system? This question can be addressed at the level of individual behavior, in terms of being an example for conventional consumers, and at the level of social forces, in terms of having an effect on the organization of food systems. As Goodman and Dupuis (2002) note, although organic food consumption is not based on a formal social movement, the philosophies of these consumers appear to constitute a vital force in society. Therefore, the food philosophies can help to interpret societal trends and contextualize ongoing developments. Most importantly, the food philosophies were associated in a theoretically meaningful way with a number of practices that are considered more sustainable than conventional ones, namely the moderate consumption of meat, the choice for seasonal and organic ingredients and the use of less processed and fresh products (Carlsson-Kanyama and González 2009; Thøgersen 2010). This linkage may provide significant cultural leverages—that is, values that motivate people to express concern about environmental and social problems and invite them to adopt more environmentally friendly lifestyles (Crompton 2011).

More specifically, there are at least four leverages that should be mentioned in the Dutch context. The first is cultivating the value of connectedness with nature. The second is cultivating the relationship between awareness and wellness. The third refers to increasing the transparency of moral aspects that are hidden in many food choices. And the fourth is shaping and supporting social norms that reflect the intrinsic value of temperance. In what follows, we discuss some examples of how these leverages could be applied.

Feeling connected with nature contributes to a feeling of responsibility and care for other creatures and the natural environment (Taylor 1989). In the context of making more sustainable food choices, connectedness with nature is a value that needs strengthening, for example, in the context of urban development. Examples of how this can be done are the development of urban agriculture to enable cities to feed themselves from within or from its neighboring communities (Dixon et al. 2009; Morgan and Sonnino 2010). Various big cities, such as New York and London, are already working on food strategies for the future. Trying to localize food production, wherever feasible, is an important component of these strategies (Morgan and Sonnino 2010). Also, new supermarket concepts that experiment with growing their products on site are interesting in this regard. More generally, initiatives that strengthen people's knowledge about the multiple links between food and nature, planting, harvesting, and preparation may serve to increase a feeling of connectedness and they are also in line with the wish for a more natural, self-determined way of living that was expressed in the Reform movements.

Second, we discussed the value of awareness. As the interviews illustrated, participants experienced independence and self-sufficiency, because they felt they could rely on their personal judgment regarding what is good to eat. This autonomy and the feeling of awareness itself were perceived as satisfying, also because participants felt that they were making choices in line with their personal values. By relying on their intuition and personal values, they felt less prone to external sources of influence, such as advertising. In terms of Taylor's framework, awareness is a crucial part of the expressive worldview, because it is a means to connect with inner/outer nature as a source of morality. Policy makers should acknowledge that this expressivity is a fundamental characteristic of Western culture that also pervades people's relationship with food (Delind 2006). They may profit from this fact by communicating about often implicit underlying values associated with more sustainable food consumption.

Third, we discussed purity as a way of living a more meaningful, moral life (Campbell 2007; Hamilton et al. 1995). The critical, idealistic approach of organic consumers has stimulated the development of environmentally relevant certification and labeling systems, which exerts continuous pressure on producers to raise sustainability standards of their production and supply chains (de Boer 2003; Lewis et al. 2010). Labeling efforts have also served to delineate between conventional and organic standards, providing a visual prompt to facilitate the purchase of more responsible products among a larger group of consumers (Morris and Winter 1999). These labels demonstrate the salience of appealing to moral motives held by a core group in society, increasing the number of people that can make more responsible choices with less effort on their part.

Fourthly, against the background of the organic philosophy, the need for personal behavior change can more easily be acknowledged and achieved. An important part of the Reform movement was about people's capacity for moral self-improvement as a practice of self-determination (Barlösius 1997). Temperance, the consumption of pure foods, and abstinence from meat were all ways in which Reformers practiced their moral values. As the interviews illustrated, these practices are still in use today (de Boer et al. 2007). Policy makers may implicitly or explicitly support social norms that reflect the intrinsic value of temperance. This could be done, for example, by promoting the consumption of large amounts of meat as normatively unacceptable.

1.5 CONCLUSION

In this study, we have made an exploratory effort to contribute to a better understanding of the cultural context of organic consumption. We have done this by trying to combine two levels of analysis: on the one hand, a top-down perspective on long-term developments in Western culture; on the other, a bottom-up perspective on contemporary organic consumers' practices, values, and beliefs. We have identified some important themes relevant to organic consumers today, and we have shown how these are rooted in a typically Western cultural background.

Organic consumption is interesting from the perspective of more sustainable food choices. Despite controversies regarding the expansion of organic production, the organic movement as a whole can be seen as a valuable driving force that stimulates the continuous improvement of food quality and inspires consumers to adopt new values and ideals that can give direction to more sustainable food practices. The feeling of connectedness with nature, awareness and purity are values that can be strengthened culturally in relation to food.

FOOTNOTES

[1]Promotion and prevention are key concepts of Higgins's psychological motivation theory (Higgins 1997), generally, a promotion orientation makes the person sensitive to gains, accomplishments, and advancement needs. In contrast, a prevention orientation makes the person sensitive to safety, responsibility, and security needs.

[2]We refer here to stores that sell the majority of their goods with organic, fair-trade or bio-dynamic certification. They typically also store Japanese foods and health food supplements.

REFERENCES

1. Aertsens, J., Verbeke, W., Mondelaers, K., & van Huylenbroeck, G. (2009). Personal determinants of organic food consumption: A review. British Food Journal, 111(10), 1140–1167.
2. Aiking, H. (2011). Future protein supply. Trends in Food Science & Technology, 22, 112–120.
3. Alimentarius, Codex. (1999). Guidelines for the production, processing, labelling and marketing of organically produced foods. Rome: FAO.
4. Badgley, C., Moghtader, J., Quintero, E., Zakem, E., Chappell, M. J., Avilés-Vázquez, K., et al. (2007). Organic agriculture and the global food supply. Renewable Agriculture and Food Systems, 22(02), 86–108.
5. Badgley, C., & Perfecto, I. (2007). Can organic agriculture feed the world? Renewable Agriculture and Food Systems, 22(02), 80–86.

6. Bakker, J. (2011). Monitor Duurzaam Voedsel 2010. Ministry of Economic Affairs, Agriculture and Innovation, The Hague [in Dutch].
7. Barlösius, E. (1997). Naturgemäße Lebensführung: Zur Geschichte der Lebensreform um die Jahrhundertwende. Frankfurt/Main: Campus Verlag GmbH.
8. Brown, K., & Kasser, T. (2005). Are psychological and ecological well-being compatible? The role of values, mindfulness, and lifestyle. Social Indicators Research, 74(2), 349–368.
9. Buck, D., Getz, C., & Guthman, J. (1997). From farm to table: The organic vegetable commodity chain of Northern California. Sociologia Ruralis, 37(1), 3–20.
10. Campbell, C. (2007). The easternization of the West. A thematic account of cultural change in the Modern era: Paradigm Publishers.
11. Carlsson-Kanyama, A., & González, A. D. (2009). Potential contributions of food consumption patterns to climate change. The American Journal of Clinical Nutrition, 89(5), 1704S–1709S.
12. Carson, R. (1962). Silent spring. Boston: Houghton Mifflin Company.
13. Charmaz, K. (2006). Constructing grounded theory: A practical guide through qualitative analysis. London: Sage Publications Ltd.
14. Constance, D. H., Choi, J. Y., & Lyke-Ho-Gland, H. (2008). Conventionalization, bifurcation, and quality of life: A look at certified and non-certified organic farmers in Texas. Southern Rural Sociology, 23, 208–234.
15. Crompton, T. (2011). Finding cultural values that can transform the climate change debate. Solutions Journal, 2(4), 56–63.
16. de Boer, J. (2003). Sustainability labelling schemes: The logic of their claims and their functions for stakeholders. Business Strategy and the Environment, 12(4), 254–264.
17. de Boer, J., Boersema, J. J., & Aiking, H. (2009). Consumers' motivational associations favoring free-range meat or less meat. Ecological Economics, 68(3), 850–860.
18. de Boer, J., Hoogland, C. T., & Boersema, J. J. (2007). Towards more sustainable food choices: Value priorities and motivational orientations. Food Quality and Preference, 18(7), 985–996.
19. de Vries, B. J. M., & Petersen, A. C. (2009). Conceptualizing sustainable development: An assessment methodology connecting values, knowledge, worldviews and scenarios. Ecological Economics, 68(4), 1006–1019.
20. de Wit, J., & Verhoog, H. (2007). Organic values and the conventionalization of organic agriculture. NJAS-Wageningen Journal of Life Sciences, 54, 449–462.
21. Delind, L. (2006). Of bodies, place, and culture: Re-situating local food. Journal of Agricultural and Environmental Ethics, 19(2), 121–146.
22. Dixon, J. M., Donati, K. J., Pike, L. L., & Hattersley, L. (2009). Functional foods and urban agriculture: Two responses to climate change-related food insecurity. New South Wales Public Health Bulletin, 20(2), 14–18.
23. Douglas, M. (1966). Purity and danger; an analysis of the concepts of pollution and taboo. London: Routledge and Kegan Paul.
24. Fiddes, N. (1991). Meat. A natural symbol. London: Routledge.
25. Fischler, C. (1988). Food, self and identity. Social Science Information, 27, 275–292.
26. Foss, D. A., & Larkin, R. W. (1976). From "The gates of Eden" to "Day of the locust." Theory and Society, 3(1), 45–64.

27. Glacken, C. J. (1967). Traces on the Rhodian shore; nature and culture in western thought from ancient times to the end of the 18th century. Berkeley, CA: University of California Press.
28. Glaser, B. G., & Strauss, A. L. (2009). The discovery of grounded theory (Vol. 4). New Jersey: Transaction Publishers.
29. Goodman, D., & DuPuis, E. M. (2002). Knowing food and growing food: Beyond the production–consumption debate in the sociology of agriculture. Sociologia Ruralis, 42(1), 5–22.
30. Gusfield, J. R. (1992). Nature's body and the metaphors of food. In M. Lamont & M. Fournier (Eds.), Cultivating differences: Symbolic boundaries and the making of inequality (pp. 75–103). Chicago: The University of Chicago Press.
31. Hamilton, M., Waddington, P. A. J., Gregory, S., & Walker, A. (1995). Eat, drink and be saved: The spiritual significance of alternative diets. Social Compass, 42(4), 497–511.
32. Hedlund-de Witt, A. (2011). The rising culture and worldview of contemporary spirituality: A sociological study of potentials and pitfalls for sustainable development. Ecological Economics, 70(6), 1057–1065.
33. Higgins, E. T. (1997). Beyond pleasure and pain. American Psychologist, 52, 20.
34. Hodgson, J. A., Kunin, W. E., Thomas, C. D., Benton, T. G., & Gabriel, D. (2010). Comparing organic farming and land sparing: Optimizing yield and butterfly populations at a landscape scale. Ecology Letters, 13(11), 1358–1367.
35. Hughner, R. S., McDonagh, P., Prothero, A., Shultz, C. J., & Stanton, J. (2007). Who are organic food consumers? A compilation and review of why people purchase organic food. Journal of Consumer Behaviour, 6(2–3), 94–110.
36. Hyland, M. E., Wheeler, P., Kamble, S., & Masters, K. S. (2010). A sense of special connection, self-transcendent values and a common factor for religious and non-religious spirituality. Archive for the Psychology of Religion/Archiv für Religionspsychologie, 32, 293–326.
37. IFOAM. (2011). The world of organic agriculture -statistics and emerging trends 2011. Bonn, Frick: IFOAM, FiBL.
38. Jackson, T. (2005). Live better by consuming less?: Is there a "Double Dividend" in sustainable consumption? Journal of Industrial Ecology, 9(1–2), 19–36.
39. Jacob, J., Jovic, E., & Brinkerhoff, M. (2009). Personal and planetary well-being: Mindfulness meditation, pro-environmental behavior and personal quality of life in a survey from the social justice and ecological sustainability movement. Social Indicators Research, 93(2), 275–294.
40. Jamison, A. (2003). The making of green knowledge: The contribution from activism. Futures, 35(7), 703–716.
41. Kass, L. R. (1994). The hungry soul: Eating and the perfecting of our nature. New York: The Free Press.
42. Kirchmann, H., Thorvaldsson, G., Bergström, L., Gerzabek, M., Andrén, O., Eriksson, L. O. (2008). Fundamentals of organic agriculture—past and present. In H. Kirchmann & L. Bergström (Eds.), Organic crop production—ambitions and limitations (pp. 13–37). The Netherlands: Springer.
43. Klintman, M., & Boström, M. (2004). Framings of science and ideology: Organic food labelling in the US and Sweden. Environmental Politics, 13(3), 612–634.
44. Knippenberg, H. (1998). Secularization in The Netherlands in its historical and geographical dimensions. GeoJournal, 45(3), 209–220.

45. Kristiansen, P., & Merfield, C. (2006). Overview of organic agriculture. In P. Kristiansen, A. Taji, & J. Reganold (Eds.), Organic agriculture: A global perspective (pp. 1–23). Collingwood, Australia: CSIRO Publishing.
46. Lang, T., & Heasman, M. (2004). Food wars. London: Earthscan.
47. Lewis, K. A., Tzilivakis, J., Warner, D., Green, A., McGeevor, K., & MacMillan, T. (2010). Effective approaches to environmental labelling of food products. Appendix A: Literature review report. London: Department for Environment, Food and Rural Affairs (Defra).
48. Montanari, M. (2006). Food is culture. New York: Columbia University Press.
49. Morgan, K., & Sonnino, R. (2010). The urban foodscape: World cities and the new food equation. Cambridge Journal of Regions, Economy and Society, 3(2), 209–224.
50. Morris, C., & Winter, M. (1999). Integrated farming systems: The third way for European agriculture? Land Use Policy, 16, 193–205.
51. Naugle, D. K. (2002). Worldview: The history of a concept. Cambridge: Wm. B. Eerdmans Publishing Co.
52. Pilgrim, S., & Pretty, J. N. (2010). Nature and culture: An introduction. In S. Pilgrim & J. N. Pretty (Eds.), Nature and culture. Rebuilding lost connections. London: Earthscan.
53. Raynolds, L. T. (2004). The globalization of organic agro-food networks. World Development, 32(5), 725–743.
54. Reijnders, L., & Soret, S. (2003). Quantification of the environmental impact of different dietary protein choices. The American Journal of Clinical Nutrition, 78(3), 664–668.
55. Rozin, P., Markwith, M., & Stoess, C. (1997). Moralization and becoming a vegetarian: The transformation of preferences into values and the recruitment of disgust. Psychological Science, 8(2), 67–73.
56. Scherr, S. J., & Sthapit, S. (2009). Mitigating climate change through food and land use. Washington: Worldwatch Institute.
57. Schifferstein, H. N. J., & Oude Kamphuis, P. A. M. (1998). Health-related determinants of organic food consumption in The Netherlands. Food Quality and Preference, 9, 119–133.
58. Schultz, P. W., Shriver, C., Tabanico, J. J., & Khazian, A. M. (2004). Implicit connections with nature. Journal of Environmental Psychology, 24(1), 31–42.
59. Small, M. L. (2009). How many cases do I need? Ethnography, 10(1), 5–38.
60. Stehfest, E., Bouwman, L., van Vuuren, D., den Elzen, M., Eickhout, B., & Kabat, P. (2009). Climate benefits of changing diet. Climatic Change, 95(1), 83–102.
61. Taylor, C. (1971). Interpretation and the sciences of man. The Review of Metaphysics, 25(1), 3–51.
62. Taylor, C. (1989). Sources of the self: The making of the modern identity. Cambridge: Harvard University Press.
63. Thøgersen, J. (2010). Country differences in sustainable consumption: The case of organic food. Journal of Macromarketing, 30(2), 171–185.
64. Tilman, D., Fargione, J., Wolff, B., D'Antonio, C., Dobson, A., Howarth, R., et al. (2001). Forecasting agriculturally driven global environmental change. Science, 292(5515), 281–284.
65. van Otterloo, A. H. (1983). De herleving van de beweging voor natuurlijk en gezond voedsel. Sociologisch Tijdschrift, 10, 507–545.
66. van Otterloo, A. H. (1999). Selfspirituality and the body: New age centres in The Netherlands since the 1960s. Social Compass, 46(2), 191–202.

67. Verdonk, D. J. (2009). Het dierloze gerecht: Een vegetarische geschiedenis van Nederland. Amsterdam: Uitgeverij Boom.
68. Versteeg, P. (2007). Spirituality on the margin of the church: Christian spiritual centres in The Netherlands. In K. Flanagan & P. C. Jupp (Eds.), A sociology of spirituality. Hampshire, England: Ashgate Publishing Limited.
69. Vogt, G. (2007). The origins of organic farming. In W. Lockeretz (Ed.), Organic farming: An international history. Oxfordshire: CAB International.
70. Weiss, R. (1994). Learning from strangers. The art and method of qualitative interview studies. New York: The Free Press.
71. Wier, M., O'Doherty Jensen, K., Andersen, L. M., & Millock, K. (2008). The character of demand in mature organic food markets: Great Britain and Denmark compared. Food Policy, 33(5), 406–421.
72. Yin, R. K. (2003). Applications of case study research. California: Sage.

CHAPTER 2

Organic Farming: The Arrival and Uptake of the Dissident Agriculture Meme in Australia

John Paull

2.1 INTRODUCTION

Four years elapsed between the coining of the term 'organic farming' and the founding of the world's first association dedicated specifically to the advocacy of organic farming (Paull, 2008). The term 'organic farming' was coined by Lord Northbourne and it first appeared in his manifesto of organic agriculture, *Look to the Land*, published in London in May 1940 (Figure 2.1). The book, published in the early days of World War II, introduced the world to not just the term 'organic farming' but also to its rationale and philosophy.

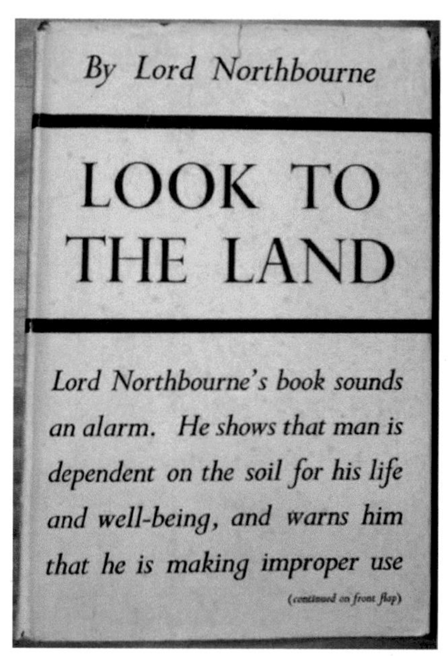

FIGURE 2.1 *Look to the Land,* Lord Northbourne, London, May 1940.

A surprising and unexplained event is how and why a term coined in wartime England (WWII) took root half a world away. The Australian Organic Farming and Gardening Society (AOFGS) was founded in Sydney in October 1944. It was the first society in the world to style itself as an 'organic' association. It was the first association to publish an 'organic' farming periodical (*Organic Farming Digest*) (Figure 4). The AOFGS was also the first association in the world to develop a set principles of 'organic farming (Paull, 2008). At the time of writing, no archives of the AOFGS have been located.

Northbourne was well aware that a differentiated agriculture needed a distinctive name and that it needed advocacy. He framed "organic farming" as a dissident agriculture in contestation with the prevailing "chemical agriculture" of his day. For Northbourne there was a battle of agricultural philosophies: "organic versus chemical farming" (1940, p.81). Northbourne had been impressed with Ehrenfried Pfeiffer's 1938 book *Bio-Dynamic Farming and Gardening* (Figure 2). He travelled to Switzerland to urge Pfeiffer to present a conference on biodynamic agriculture to a British audience. Pfeiffer was, at the time, the leading advocate of biodynamics. Pfeiffer had 'outed' biodynamics to a broad audience by publishing Bio-Dynamic Farming and Gardening (1938) (Paull, 2011c). An outcome of Northbourne's visit to Pfeiffer was the Betteshanger Summer School and Conference on Bio-Dynamics Farming, which Northbourne hosted at his farm in Kent and at which Pfeiffer was the key presenter (Northbourne, 1939; Paull, 2011b). Within months of the Betteshanger Biodynamics Conference, Britain and Germany were at war, so the window of opportunity for successfully spruiking Germanic agricultural ideas was slammed shut (Pfeiffer was German, Rudolf Steiner was Austrian). Northbourne reframed the Steiner/Pfeiffer call for an agriculture free of synthetic inputs and in harmony with the cosmos for an Anglo audience. He stripped out the overt mystical and anthroposophic trimmings and distanced its Steinerian provenance.

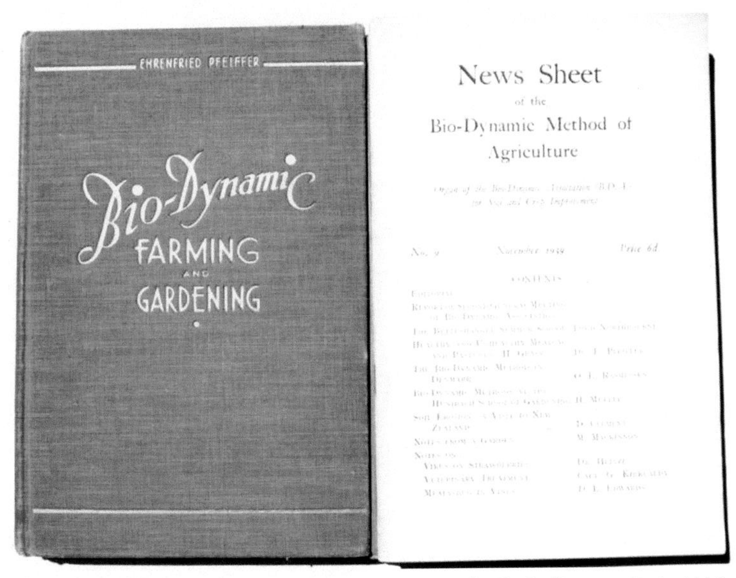

FIGURE 2.2 *Bio-dynamic Farming and Gardening*, Ehrenfried Pfeiffer, New York, 1938.

Look to the Land was an organics manifesto that Northbourne was temperamentally and experientially ideally positioned to write. He was a gifted wordsmith, a visionary thinker, a spiritually grounded individual, a graduate and lecturer in agriculture of Oxford University, the Chairman of Swanley Horticultural College, a Governor of the agricultural Wye College, as well as an experienced farmer. Look to the Land presented the rationale for an agriculture that was an alternative to chemical farming and which he dubbed 'organic farming'.

After the Betteshanger Conference, Pfeiffer migrated to the USA and was mentor to Jerome Rodale. In titling his book, *Bio-Dynamic Farming and Gardening*,

Pfeiffer coupled 'farming' with 'gardening'. Rodale followed Pfeiffer's lead in titling his own periodical *Organic Farming and Gardening* (Figure 2.3). Although Northbourne was himself a keen gardener and he wrote of "our national love of gardening" and that "Our love of gardening can blossom into something greater" (1940, p.107) he nevertheless did not himself couple the terms 'farming and gardening' in his book. In this, Northbourne did not follow Steiner's lead (Steiner founded the Experimental Circle of Anthroposophic Farmers and Gardeners in 1924) however he did cite Pfeiffer's Bio-Bibliography" (p.196).

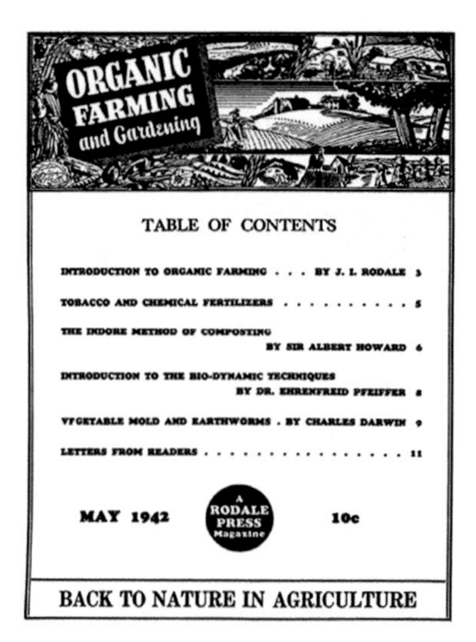

ORGANIC FARMING and Gardening

TABLE OF CONTENTS

INTRODUCTION TO ORGANIC FARMING . . . BY J. I. RODALE 3

TOBACCO AND CHEMICAL FERTILIZERS 5

THE INDORE METHOD OF COMPOSTING BY SIR ALBERT HOWARD 6

INTRODUCTION TO THE BIO-DYNAMIC TECHNIQUES BY DR. EHRENFREID PFEIFFER 8

VEGETABLE MOLD AND EARTHWORMS . BY CHARLES DARWIN 9

LETTERS FROM READERS 11

MAY 1942 — A RODALE PRESS Magazine — 10c

BACK TO NATURE IN AGRICULTURE

FIGURE 2.3 *Organic Farming and Gardening,* Rodale Press, USA, May 1942.

Jerome Rodale was an early adopter of Northbourne's 'organic farming' meme. Rodale was a publisher with a record of harvesting British material and appropriating, repurposing and repackaging it for an American audience (Jackson, 1974; Rodale,1965). The first issue of his periodical Organic Farming and Gardening was dated May 1942. It was the world's first periodical devoted to the advocacy of organic agriculture and the timing places it as a candidate for influencing the establishment of the AOFGS.

Rodale was mentored by Pfeiffer. Pfeiffer had migrated from Switzerland to the USA (Selawry, 1992). Pfeiffer's biodynamics book introduced a global audience to Rudolf Steiner's biodynamic agriculture. The book fulfilled Steiner's injunction, of his Agriculture Course presented at Koberwitz (now

Kobierzyce, Poland) to put his "hints" to the test and develop them to a form suitable for publication (Steiner, 1924). Pfeiffer's book appeared in five languages, English, French, Italian, German and Dutch (1938a, 1938b, 1938c, 1938d, 1938e).

Rodale and the AOFGS were early adopters of Northbourne's 'organic' terminology and in this they were ahead of their contemporaries. Eve Balfour and Albert Howard, for example, and other contemporary authors on kindred themes were slower in the uptake. Eve Balfour quoted Northbourne extensively in her book *The Living Soil* (1943) with pages 14 to 17 of her book being a lengthy direct quote from *Look to the Land* (although it is barely differentiated in her text as a quotation and a reader may easily miss the, incorrectly dated, attribution to Northbourne). Balfour's book did not include a mention of 'organic' farming or agriculture and when the Soil Association was founded in London, the Memorandum and Articles of Association (Douglas, 1946) made no mention of 'organic'. Similarly, Albert Howard did not use the term 'organic farming' or derivatives in his books, including *Farming and Gardening for Health and Disease* (Howard & Howard, 1945). Howard's book was republished as *The Soil and Health: Farming and Gardening for Health and Disease* (Howard, 1945b) and has more recently been oddly retitled by the University Press of Kentucky as *The Soil and Health: A Study of Organic Agriculture* (Howard, 1945a) despite the fact that Howard's text makes no mention of 'organic' farming or agriculture.

The AOFGS was founded in October 1944. After a decade of national organics advocacy it was wound up on 19 January 1955. The key vehicle of advocacy for the AOFGS as well as the major expense of the society was their periodical, the *Organic Farming Digest*. The first issue appeared in April 1946, which was just as soon as wartime restrictions on paper supplies were lifted finally enabling publication. The Society however failed to find a viable business model and the financial strain of publishing a periodical eventually led to the demise of the Society. The final issue of the periodical, by then titled *Farm & Garden Digest* (incorporating *Organic Farming Digest*), was the 29th issue and dated December 1954. The Society was wound up at a meeting of 19 January, 1955 at the Primary Producers' Union Office, Sydney (Paull, 2008).

The present paper investigates the arrival, reception and uptake in Australia of the organic farming meme and seeks to distinguish between two potential candidates, Northbourne (1940) and Rodale (1942), as the genesis of the AOFGS, and to determine whether Pfeiffer (1938) perhaps also played a role.

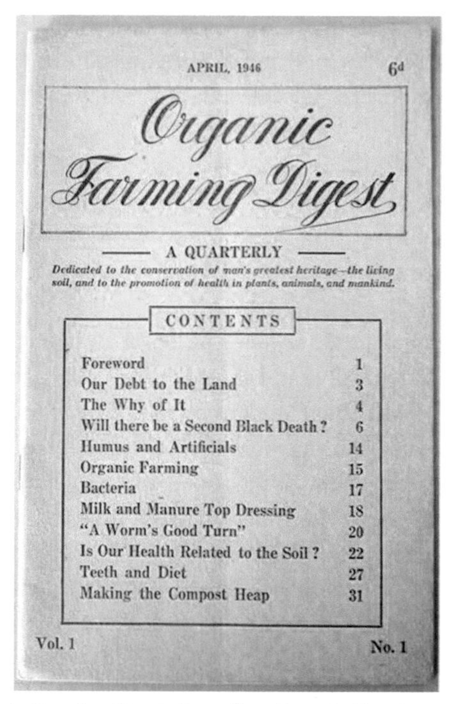

APRIL, 1946 6d

Organic Farming Digest

A QUARTERLY

Dedicated to the conservation of man's greatest heritage—the living soil, and to the promotion of health in plants, animals, and mankind.

CONTENTS

Foreword	1
Our Debt to the Land	3
The Why of It	4
Will there be a Second Black Death?	6
Humus and Artificials	14
Organic Farming	15
Bacteria	17
Milk and Manure Top Dressing	18
"A Worm's Good Turn"	20
Is Our Health Related to the Soil?	22
Teeth and Diet	27
Making the Compost Heap	31

Vol. 1 No. 1

FIGURE 2.4 *Organic Farming Digest,* Australian Organic Farming and Gardening Society, Sydney, April 1946

2.2 METHODOLOGY

The National Library of Australia (NLA) maintains the largest archive of Australian publications and newspapers. The online data base of the NLA, Trove (trove.nla.gov.au), includes 882 digitised newspapers (NLA, 2015). The NLA digitised newspapers are from Australian Capital Territory (n=8), New South Wales (n=324), Northern Territory (n=8), Queensland (n=66), South Australia (n=59), Tasmania (n=43), Victoria (n=323), Western Australia (n=49), and National (n=2). The NLA database of digitised titles is the primary source of material for the present research. No archive of the Australian Organic Farming and Gardening Society (AOFGS) was located and that is consistent with previous research (Jones, 2010; Paull, 2008)

The NLA newspapers were searched for the period from 1 January 1938 up to 19 January 1955 (these dates were chosen because Pfeiffer's book appeared in 1938 and the AOFGS was wound up on 19/1/1955). The results were sorted into two periods, pre- AOFGS and after the founding of the AOFGS; this was operationalised as pre 14 October 1944 and post 14 October 1944 on the grounds that the first identified public appearance of the AOFGS was on this date (Jeremy, 1944) and the Society was wound up on 19 January 1955 (Paull, 2008). The newspapers of all Australian states and territories were searched (viz.: National, ACT, NSW, NT; Qld; SA; Tas; Vic; and WA). All article categories were searched (viz.: Article; Advertising; Detailed Lists, Results, Guides; Family Notices; and Literature). All article lengths were searched. Searches were not case sensitive.

Items searched:

(a) Pfeiffer mentions: items mentioning Ehrenfried Pfeiffer and/or his book Bio-Dynamic
(b) Farming and Gardening (1938a) and/or bio-dynamic and/or biodynamic;
(c) Northbourne mentions: items mentioning Lord Northbourne and/or his book Look to the Land (1940). Mentions of the social or political life of Lord Northbourne were excluded;
(d) Rodale mentions: items mentioning Jerome Rodale and/or Rodale Press and/or his periodical Organic Farming and Gardening (1942); and
(e) Organic Farming mentions: items mentioning 'organic farming' and/ or organic agriculture' and/or derivative terms viz. 'organic farm' and 'organic farmer'.

2.3 RESULTS

Prior to the founding of the AOFGS (i.e. pre 14 October 1944) there were no Organic Farming mentions in the Australian press (Table 1, Figure 5). There were no Rodale mentions (Table 2.1). There were two Pfeiffer mentions (viz. Cairns Post, 1942; *Queensland Country Life*, 1939) (Table 1). There were 14 Northbourne mentions and these included reviews, articles, and advertisements for the book (Tables 1 & 2). Northbourne mentions appeared in the press in four states (NSW, n=4; SA, n=6; Qld, n=1; WA, n=3) from July 1940 to September 1944, i.e. up to a month prior to the launch of the AOFGS (Table 2.2).

In the period of the life of the AOFGS (i.e. from 14 October 1944 to 19 January 1955) there were 25 Pfeiffer mentions, 9 Northbourne mentions, 9 Rodale mentions, and 353 Organic Farming mentions (Table 2.1, Figure 2.5).

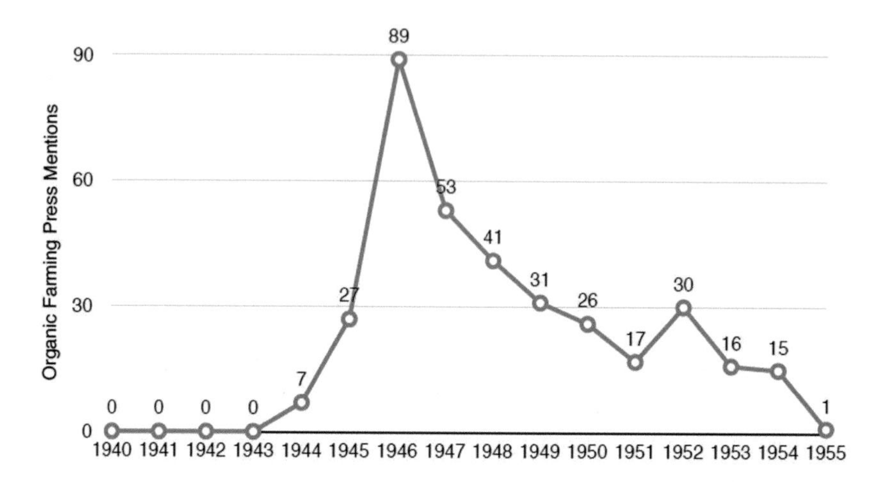

FIGURE 2.5 Number of mentions of Organic Farming in the Australian press before and during the life of the AOFGS.

TABLE 2.1 Mentions of Pfeiffer, Northbourne, Rodale, and Organic Farming in the Australian press prior to the founding of the AOFGS and during the life of the AOFGS.

Items	Press mentions prior to AOFGS	Press mentions during AOFGS
(a) Pfeiffer mentions	2	25
(b) Northbourne mentions	14	9
(c) Rodale mentions	0	9
(d) Organic Farming mentions	0	353

Within two months of its publication in London, Northbourne's *Look to the Land* was available in Australia (Table 2.2). The first mention of *Look to the Land* in the Australian press was the announcement of "Books Received" on 13 July 1940 in the leading newspaper of South Australia (Advertiser, 1940a). A favourable review of *Look to the Land* appeared the following month (Advertiser, 1940b) (Table 2).

TABLE 2.2 Mentions of Northbourne in the Australian press prior to the founding of the Australian Organic Farming and Gardening Society (AOFGS).

Date	Title	Newspaper, State
Look to the Land published in London, May 1940		
13 July 1940	Books Received: General (Advertiser, 1940a)	The Advertiser, SA
24 August 1940	The Land Problem (Advertiser, 1940b)	The Advertiser, SA
4 December 1940	Man and the Soil (Macleay Chronicle, 1940)	The Macleay Chronicle, NSW
21 September 1940	Land-Love (SMH, 1940)	The Sydney Morning Herald, NSW
4 December 1940	Man and the Soil (Macleay Chronicle, 1940)	The Macleay Chronicle, NSW
2 August 1941	Rigby's for Books (Rigby Ltd., 1941)	The Advertiser, SA
6 September 1941	New Books and Publications (Alberts Bookshop, 1941)	The West Australian, WA
8 November 1941	Books of the Week: "Look to the Land" by Lord Northbourne (Palette, 1941)	News, SA
22 November 1941	Life & Letters (West Australian, 1941)	The West Australian, WA
27 November 1941	The Farm Bookshelf: When The Soil Dies (Martingale, 1941)	The Western Mail,WA
25 March 1943	Soil and Humus (Mount Barker Courier, 1943)	The Mount Barker Courier & Onkaparinga & Gumeracha Advertiser, SA
6 January 1944	Food Front Plan (Queensland Country Life, 1944)	Queensland Country Life, Qld
3 February 1944	Nemo's Horticultural Notes (Nemo, 1944)	The Murray Pioneer, SA
16 September 1944	The Rule of Return (Farmers' Friend, 1944)	Northern Star, NSW

Australian Organic Farming and Gardening Society (AOFGS) founded c.14 October 1944

Northbourne's book was well received in the Australian press. Reviews were all favourable. They appear to have drawn most of their material from Chapter 1. None of the accounts of Northbourne's book mention his core tenet of 'organic farming' and none mention his framing of the agricultural contest of the times as 'organic versus chemical farming'. Nevertheless, the early Australian accounts of Northbourne's book were thoughtful reviews which retain their salience to this day and make for interesting reading; a selection follows.

The first review of *Look to the Land* in the Australian press reported:

> Lord Northbourne attacks the subject of the biological sickness of the world by saying that the economic and spiritual sicknesses of the world are aspects of one and the same phenomenon. As agriculture is the basis of man's existence on this planet, so must agriculture depend on the soil.
>
> He then discusses the properties and potentialities of the soil, emphasising the fact that soil is 'living.' Any discussion of this kind naturally leads to the problem of soil erosion, which is set out in the light of alarming statistics from America and interesting facts from China. Lord Northbourne blames the folly of those who, in their endeavour to 'get rich quickly,' exhaust the land on which they depend.
>
> He complains that our present financial system is partly responsible for the evils he deplores; he says that it tends to drive people to the cities, and is placing intolerable burdens on future generations. The remedy suggested for soil erosion is a strict adherence to 'the rule of returns.' What has been taken from the land must be put back in some form, if the fertility of the soil is to be maintained.
>
> Lord Northbourne has little faith in social legislation or the activities of great public bodies: he insists, however, on the necessity for a change of heart in the individual and a realisation of the importance of the family as a social unit. He wants to see families returning to 'the good earth' and settled on self-contained farms.
>
> Lord Northbourne is himself a considerable landowner and has had farming experience, so his idealism is not that of an unpractical theorist. He wants to develop a more spiritual outlook, a greater strength of character, and a deeper understanding of Nature than is now apparent in our overgrown urbanised populations (*Advertiser,* 1940b, p.10).

The first advertisement in the Australian press for *Look to the Land* stated:

Lord Northbourne suggests that present commercial and industrial tendencies particularly in their effects on farming cannot be allowed to continue to spread disease and disorder, nor to reduce world fertility to starvation point. Price 9/9 (Rigby Ltd., 1941, p.10).

Look to the Land featured as book of the week in *The News* where it was recommended as "a book for everyone":

> Lord Northbourne is a practical farmer, there is not much of the social or big business men about him. He is a large land owner and makes use of every yard of it in mixed farming and market gardens, working as hard as any employee, and, judging by his forthright character, as exemplified in this book, is a man after our own hearts and worth listening to.
>
> He shows that man is dependent on the soil for his life and well being, and warns us all that we are not making proper use of our heritage. He tells you all about it in plain, everyday language, and makes his plaint so interesting and understandable that even the scribe who writes this covered the contents twice over.
>
> ... Starting off with the nutritional needs of plants, the importance of humus, he goes on to that dread subject, with which we in Australia can sympathise and under stand—erosion.
>
> These chapters on erosion are truly frightening, and are not the least overdrawn or exaggerated ...The Missouri basin has lost an average of seven inches of top soil in 24 years, which sounds pretty dreadful when learned professors have estimated the mean rate of soil formation as one inch in 10,000 years.
>
> According to the author, much the same is true of many other countries. He makes the flat announcement that "Australia is going faster than America," but has only been under 'civilised' Influence for one-third of the time of the United States. He gives his reason, overstocking and unsound cultural methods.
>
> Well, I leave it to those who are in a better position to know as to whether Lord Northbourne is right or not, but, judging from what has been written and illustrated in books, papers, and magazines, I take it that this noble farmer's arguments will be fairly solid hurdles to cover.
>
> Lord Northbourne discusses the farmer, and says he has a reputation for individualism and independence. "These are sound qualities, but they are not appreciated in modern, large-scale business," he says, adding, "But that is a debased form of organisation."
>
> ... He holds that the farmer is not incompatible with the highest form of social life, and gives it value, for he improves the quality of the smallest units, from which any such organisation must grow.
>
> In reading Lord Northbourne, I feel sure that he has had a thorough education in every phase of life, and he said one thing that was driven into my nut, when a young art student ... That is:-"If we want to succeed in the great task before us we must adopt a humbler attitude towards the elementary things of life than

that which is implied in our frequent boasting about our so-called 'Conquest of Nature.' We have put ourselves on a pinnacle in the pride of an imagined conquest. It is just as sensible as if a man should try to cut off his own head, so as to isolate his superior faculties."

A book for everyone, be he parson, docker, clerk, or farmer, and especially the suburban gardener (Palette, 1941, p.2).

The *Sydney Morning Herald's* review of *Look to the Land*:

Lord Northbourne is himself a farmer, but instead of being a strictly practical advocacy of sane agriculture, as one might expect, his book turns out to be an almost mystical Interpretation of the relation of man to earth and a plea for recognition of more than material considerations In farming. Somewhat after the style of Meredith, he shows how love binds all creatures and the earth together and that any disturbance of this union for gain creates a profound unsettlement In the life of man and results in the impoverishment of the soil. ... He believes man is paying for his disturbance of the normal order through greed and his general remedy is a return to the land particularly the small farm and a loving cultivation of it. Social life could well begin again from this basis, he considers the land must be studied not exploited.

His book is full of Interesting matter—discussions of economics, health, chemical farming, international relations—and it may provide valuable suggestions in a new order after the war (SMH, 1940, p.10).

The *West Australian* reported positive sales results:

Published over a year ago by Dent's, Lord Northbourne's book 'Look to the Land' is selling increasing numbers every week. (Price 7/6) (*West Australian*, 1941, p.7). The book was reviewed in Perth's *Western Mail*: Lord Northbourne is a landowner in Kent and Northumberland who runs a mixed farm and market gardens. His recently-published book, "Look to the Land" is one that should be read by all thinking men for in it he sounds an alarm which the world should heed, in man's exploitation of the soil; in his reckless wastage of his heritage, fie traces the root cause of most of the physical, social, political, and economic ills from which the world is suffering. I started to review the book as a whole but found that the first chapter contained so much food for thought that I have contented myself with presenting in a condensed form, a few of the author's pronouncements concerning soil erosion contained therein ...

Lord Northbourne, author of 'Look to the Land,' modestly describes his book as an attempt by a layman, writing for laymen, to set forth how much more there is in agriculture than the mere production of cheap and abundant food.

The problems confronting agriculture throughout the world are not merely farmers' problems, he points out, for the soil is the foundation of the physical life of man, whether he be a farmer or a city dweller. It is the background of every man's life for he is dependent upon the soil for his nutrition.

The Soil is an Entity ... soil is a complex mixture of many ingredients living and non-living; a whole world in itself and a living entity ... Increases of production, the author points out, must not be taken as increases in fertility. Actually an increase in production is usually secured by "cashing in" on existing fertility, and, as it is used up we encounter the disastrous results which have been described ...

The temptation to exploit the fertility of the soil for immediate personal gain is no new thing, but during recent years man has enormously extended his physical powers by the use of mechanical devices so that one man can now do what used to be the work of hundreds and can do it faster. With the speeding up of cultural operations there has been a speeding up of erosion ...

Man sets about his desert making in various ways. He alters the texture of the soil by using up humus and failing to replace it—by failing to feed the soil with organic matter; livestock are the great converters of otherwise unwanted organic matter to a form in which it can be used by plants. Stockless farming, under-stocking, burning straw, etc., are all cases of failure to observe the 'rule of return' which is the essence of farming. Only by faithfully returning to the soil in due course everything that has come from it can fertility be made permanent and the earth be made to yield a genuine increase.

Large-scale monoculture (the growing of one crop only) upsets the balance of factors in the soil in many ways. There is no give and take between crops. Disease spreads easily. Nature always provides a mixture of plants, and of animals; only so can living matter be kept constantly in circulation without wastage ...The injudicious felling of timber may lead to much more than denudation of the hills on which the timber grows. Forests act as sponges, and level out the rate at which water leaves the hills. Thus injudicious deforestation leads to erosion on the hills, and to alternations of flood and drought in the valleys, ending in erosion or harmful silting of the valleys themselves ...

Debt and Destruction. "The rapid extension of exhaustive farming throughout the world is linked with the roughly simultaneous extension of a peculiar economic system which has led to a vast accumulation of financial debt. Such debt both internal and international has grown to a point at which repayment is practically impossible and the mere payment of interest is severely oppressive ...

Under present conditions the only thing that pays is quick profit making whilst the going is good. By ignorant or unscrupulous exploitation and exhaustion of

> fertility- vast profits have been made (by financiers rather than by farmers) in the name of cheap food. The pace is forced for the sound farmer wherever he lives.
>
> As is usual nowadays, it will be left to future generations to pay for our mistakes, but they may not have the wherewithal. Money alone is notoriously useless in a desert.
>
> "Look to the Land," by Lord Northbourne. Published by J. M. Dent and Sons Ltd. Price 9/6 (Martingale, 1941, pp.56-7).

Nemo (1944) reviewed *Look to the Land* in the *Murray Pioneer,* a South Australian rural newspaper. In the preamble he stated that:

> Mr. W. Macgillivray (Member [of Parliament] for Chaffey), has kindly sent along a book for me to read which is so full of 'meat' that I don't know where to start to give it a full review. It Is called "Look to the Land" by Lord Northbourne and deals with every aspect of agriculture in England, in the political, economic and scientific aspect, but covers a wide range of other countries as well" (p.8).

Northbourne's book continued to attract mentions in the Australian press during the life of the AOFGS with nine mentions in that period, all favourable or neutral (Table 3).

TABLE 2.3 Northbourne mentions in the Australian press during the life of the Australian Organic Farming and Gardening Society (AOFGS).

Australian Organic Farming and Gardening Society (AOFGS) founded c.14 October 1944		
4 January 1945	Soil Fertility (Martingale, 1945)	Western Mail, WA
17 January 1945	Letters to the Editor: Soil Erosion (Daft, 1945)	The Mercury, TAS
7 July 1945	The Future Of Our Farm Industries (Scrutator, 1945)	Northern Star, NSW
13 July 1945	Books For Farmers (Alberts Bookshop, 1945)	Western Mail, WA
22 March 1946	Agricultural Bureau. Y.P Conference (Kadina & Wallaroo Times, 1946)	The Kadina and Wal-laroo Times, SA
4 July 1946	The Economics of Primary Production (Till, 1946)	Murray Pioneer, SA
13 August 1948	Rural Review: The Abuse of the Land (Cobbett, 1948)	The New Times, VIC
20 June 1950	World Soil Erosion (Mannock, 1950)	Camperdown Chronicle, VIC
18 August 1951	Preece's Farming (Preece, 1951)	The Advertiser, SA

Australian Organic Farming and Gardening Society (AOFGS) wound up 19 January 1955

No archive of the AOFGS was located. Colonel Harold White was prominent in the AOFGS, and was most likely a founder. He was the second most prolific author (after the editor, V H Kelly) of articles in the *Organic Farming Digest*, contributing 20 articles (of a total of 378 published articles) over the 29 issues of the *Digest*. White was an enthusiastic advocate of organic farming of the AOFGS, presenting lectures, doing radio interviews (e.g. White, 1954), writing letters to the editor, authoring a pamphlet (White, 1959) and co-authoring a book (White & Hicks, 1953). Of the participants in the AOFGS only White's very incomplete papers were located by the author and some of his library remains in the possession of his family. The remnants of White's library and papers revealed no copy of Northbourne's *Look to the Land*, no copy of Ehrenfried Pfeiffer's *Bio-Dynamic Farming and Gardening*, no issues of Rodale's periodical *Organic Farming and Gardening*, no books of Rudolf Steiner, and no material of the Experimental Circle of Anthroposophical Farmers and Gardeners (ECAFG). White's surviving papers, as inspected by the author, throw little or no light on the genesis of the AOFGS; a case of absence of evidence rather than evidence of absence.

The ECAFG was active in Australia from 1928 with its members practicing Anthroposophical and biodynamic farming, precursors to organic farming (Paull, 2013, 2014a, 2014b). On the evidence as it stands at present, the genesis and activities of the ECAFG and the AOFGS appear to have proceeded independently of each other.

2.3 DISCUSSION AND CONCLUSION

The AOFGS adopted the term 'organic farming' as the defining raison d'être of the Society. The present research reveals that the book in which the term 'organic farming' first appeared, Northbourne's *Look to the Land*, was available in Australia shortly after being published in London. It was advertised for sale in the four years prior to the founding of the AOFGS, as well as after. The book was favourably reviewed in the Australian press as a "Book of the Week" and it was recommended as a "book for everyone" which was "full of Interesting matter".

The evidence available indicates that the AOFGS took up the term 'organic farming' directly from Northbourne's book. *Look to the Land* was reissued in Britain in 1942 bearing the "War Economy Standard" logo and 1946. The book continued to be advertised in the Australian press suggesting that these editions made their way to Australia despite the shipping challenges of the time.

Rodale adopted Northbourne's 'organic farming' term for his US periodical *Organic Farming and Gardening* (1942). There were no mentions of Rodale nor of Organic Farming and Gardening in the Australian press prior to the founding of the AOFGS. The evidence suggests that Rodale played no role in the genesis of the AOFGS. Historically, books and periodicals have generally entered the Australian market via British distributors, rather than American, and the apparent absence of Rodale's periodical in Australia at this time is no great surprise and confirms this market distortion of the time.

Pfeiffer and his book *Bio-Dynamic Farming and Gardening* received scant attention in the Australian press prior to the founding of the AOFGS with just two mentions, and the book was neither reviewed nor advertised in the Australian press. The evidence suggests that it likely played little or no role in influencing the genesis of the AOFGS.

The present study leaves open the question of how the founders of the AOFGS settled on 'Farmers and Gardeners' in the title of their new society. The coupling of these terms—farmers and gardeners—generates a phrase that is generally absent in the agricultural publications and associations of the time (as well as now). The phrase 'Farmers and Gardeners' has nevertheless been used within the dissident agriculture movements from at least 1924. At his seminal Agriculture Course at Koberwitz, Rudolf Steiner (1924) established the Experimental Circle of Anthroposophical Farmers and Gardeners (ECAFG) (Paull, 2011a). When Pfeiffer (1938) presented the work of Steiner and the ECAFG to the world, he carried forward the phrase 'Farmers and Gardeners' in titling his book as *Bio-Dynamic Farming and Gardening*. Four years later, when Rodale (1942) adopted Northbourne's term 'organic farming', and at a time when he was taking guidance from Pfeiffer, he titled his periodical *Organic Farming and Gardening*.

A standout feature of the tally of press mentions was that prior to the founding of the AOFGS there was not a single Organic Farming mention, while, in contrast, during the life of the AOFGS there were 353 Organic Farming mentions in the Australian press (Table 2.1, Figure 2.5). This is despite the wide coverage and availability of *Look to the Land*.

In the absence of the testimony of the founders of the AOFGS (which is lacking) and/or the emergence of the archives and records of the AOFGS (which remains a possibility), conclusions must remain tentative. That said, the conclusion to be drawn is that the founders of the AOFGS took up the term 'organic farming' from Northbourne's book, and it is, after all, the core meme of the book and its foundational idea.

AOFGS archives, minutes, and/or records may at some future point come to light. Australia has the advantage over many other countries in that there has been no destruction of documents due to bombing (for example, as occurred in Britain in WWII), no wartime seizures of documents as booty or intelligence (for example, as apparently occurred with Anthroposophical Society documents in Nazi Germany and during WWII). Added to this, Australian homes are larger than those of most countries and this is a cause for optimism in recovering 'lost' documents which may still be 'tucked away' in spare rooms or sheds.

The archives of two Australian organics organisations have been secured. Records of the Living Soil Association of Tasmania (LSAT) are deposited in the State Library of Tasmania (Paull, 2009a) and the records of the Soil Association of South Australia (SASA) have more recently been deposited by the SASA archivist, Dr Sandra Grimes, into the State Library of South Australia (SLSA) (Paull, 2009b).

At the demise of the AOFGS, the *Digest* reported: "The Society has always operated under a financial handicap, and for this reason the *Digest* fell short in some respects. However the principles of organic farming have been sufficiently publicised for the work to continue ... there is solace in the fact that it has performed a service in publicising organic farming principles in Australia" (The Executive Officers, 1954, p.1).

The successful advocacy of organics by the AOFGS is clearly evidenced by the 353 mentions of Organic Farming in the Australian press during the course of the life of the AOFGS (1944-1952) and is contrasted by the zero mentions of Organic Farming prior to the founding of the AOFGS (Table 2.5). The rise and fall of the Australian press mentions of Organic Farming (Figure 2.5) is a proxy index of the organics advocacy of the AOFGS and offers a proxy snapshot of the rise and fall of the AOFGS, Australia's first organics association.

REFERENCES

1. Advertiser. (1940a). Books Received: General. Adelaide: The Advertiser. 13 July, p. 4.
2. Advertiser. (1940b). The Land Problem: 'Look to the Land' by Lord Northbourne. Adelaide: The Advertiser. 24 August, p. 10.
3. Alberts Bookshop. (1941). New books and publications. Perth, WA: The West Australian. 6 September, p.7.

4. Alberts Bookshop. (1945). Books for Farmers. Perth: Western Mail. 13 July, p.67.
5. Balfour, E. B. (1943). The Living Soil: Evidence of the importance to human health of soil vitality, with special reference to post-war planning. London: Faber and Faber.
6. Cairns Post. (1942). Primary production. Cairns, Qld.: Cairns Post. 19 September, p.4.
7. Chronicle, M. (1940). Man and the Soil. Kempsey, NSW: The Macleay Chronicle. 4 December, p.6.
8. Cobbett. (1948). The New Times. Melbourne, VIC: The New Times. 13 August, pp.10-11.
9. Daft, W. (1945). Letters to the Editor: Soil Erosion. Hobart: The Mercury, 17 January, p.14.
10. Douglas, F. C. R. (1946). Memorandum and Articles of Association of The Soil Association, Ltd. 12 pp.; London: F. C. R. Douglas, Solicitor.
11. Farmers' Friend. (1944). The Rule of Return. Lismore, NSW: Northern Star. 16 September, p.6.
12. Howard, A. (1945a). The Soil and Health: A Study of Organic Agriculture (2006, first titled thus; original title: Farming and Gardening for Health and Disease ed.). Lexington, KY: The University Press of Kentucky.
13. Howard, A. (1945b). The Soil and Health: Farming and Gardening for Health and Disease (1947; first titled thus; original title: Farming and Gardening for Health and Disease ed.). Emmaus, PA: Rodale Books.
14. Howard, A. & Howard, L. E. (1945). Farming and Gardening for Health and Disease. London: Faber and Faber.
15. Jackson, C. (1974). J. I. Rodale: Apostle of Nonconformity. New York: Pyramid Books.
16. Jeremy, E. (1944). Nation's health depends on soil health. Burnie, Tas.: The Advocate. 14 October, p.6.
17. Jones, R. (2010). Green Harvest: A History of Organic Farming and Gardening in Australia. Melbourne: CSIRO Publishing.
18. Kadina & Wallaroo Times. (1946). Agricultural Bureau. Y.P. Conference. South Australia: The Kadina and Wallaroo Times. 22 March, p.2.
19. Macleay Chronicle. (1940). Man and the Soil. Kempsey, NSW: The Macleay Chronicle. 4 December, p.6.
20. Mannock, C. M. (1950). World Soil Erosion. Victoria: The Camperdown Chronicle. 20 June, p.7.

21. Martingale. (1941). The Farm Bookshelf: When The Soil Dies. Perth: Western Mail. 27 November, p.56-7.
22. Martingale. (1945). Soil Fertility - "Groucher" Asks Some Questions. Perth: The Western Mail. 4 January, p.49.
23. Mount Barker Courier. (1943). Soil and humus. SA: The Mount Barker Courier and Onkaparinga and Gumeracha Advertiser. 25 March, p.2.
24. Nemo. (1944). Nemo's Horticultural Notes: "Look to the Land". Renmark, SA: The Murray Pioneer. 3 February, p.8.
25. NLA. (2015). Newspaper and magazine titles. Canberra: National Library of Australia. http:// trove.nla.gov.au/ndp/del/titles accessed 1/4/2015.
26. Northbourne, Lord. (1939). The Betteshanger Summer School. News Sheet of the Bio-Dynamic Method of Agriculture, 9, 8-11.
27. Northbourne, Lord. (1940). Look to the Land. London: Dent.
28. Northbourne, Lord. (1942). Look to the Land (Second impression) ("War Economy Standard"; 1942 ed.). London: Dent.
29. Northbourne, Lord. (1946). Look to the Land (Third impression) (1946 ed.). London: Dent.
30. Palette. (1941). Books of the Week: "Look to the Land" by Lord Northbourne. Adelaide: The News. 8 November, p. 2.
31. Paull, J. (2008). The lost history of organic farming in Australia. Journal of Organic Systems, 3(2), 2-17.
32. Paull, J. (2009a). The Living Soil Association: Pioneering organic farming and innovating social inclusion. Journal of Organic Systems, 4(1), 15-33.
33. Paull, J. (2009b). The Path to Otopia: an Australian Perspective: Keynote Address: Launch of the Historical Research Archive of the Soil Association of South Australia (SASA), State Library of South Australia, Adelaide, 4 February.
34. Paull
35. Journal of Organics (JO), Volume 2 Number 1, 2015
36. Paull, J. (2011a). Attending the first organic agriculture course: Rudolf Steiner's Agriculture Course at Koberwitz, 1924. European Journal of Social Sciences, 21(1), 64-70.
37. Paull, J. (2011b). The Betteshanger Summer School: Missing link between biodynamic agriculture and organic farming. Journal of Organic Systems, 6(2), 13-26.

38. Paull, J. (2011c). Biodynamic Agriculture: The journey from Koberwitz to the World, 1924-1938. Journal of Organic Systems, 6(1), 27-41.
39. Paull, J. (2013). A history of the organic agriculture movement in Australia. In B. Mascitelli & A. Lobo (Eds.), Organics in the Global Food Chain (pp. 37-60). Ballarat: Connor Court Publishing.
40. Paull, J. (2014a). Ernesto Genoni: Australia's pioneer of biodynamic agriculture. Journal of Organics, 1(1), 57-81.
41. Paull, J. (2014b). Lord Northbourne, the man who invented organic farming, a biography. Journal of Organic Systems, 9(1), 31-53.
42. Pfeiffer, E. (1938a). Bio-Dynamic Farming and Gardening: Soil Fertility Renewal and Preservation (F. Heckel, Trans.). New York: Anthroposophic Press.
43. Pfeiffer, E. (1938b). De Vruchtbaardheid der Aarde: Haar Behoud en Haar Vernieuwing het Biologisch-Dynamische Principe in de Natuur. Deventer, Netherlands: N.V. Uitgevers-Maatschappij & E. Kluwer.
44. Pfeiffer, E. (1938c). Die Fruchtbarkeit der Erde Ihre Erhaltung and Erneuerung: Das Biologisch-Dynamische Prinzip in der Natur. Basle, Switzerland: Verlag Zbinden & Hugin.
45. Pfeiffer, E. (1938d). Fécondité de la Terre, Méthode pour conserver ou rétablir la fertilité du sol: Le principe bio-dynamique dans la nature. Paris: Editions de La Science Spirituelle.
46. Pfeiffer, E. (1938e). La Fertilita della Terra come Restaurarla e come Conservarla. Milano, Italy: La Prora.
47. Preece. (1951). Preece's Farming. Adelaide, SA: The Advertiser. 18 August, p.6.
48. Queensland Country Life. (1939). Soil fertility factor in human health? Startling theory advanced by British medico. Qld: Queensland Country Life. 6 April, p.5.
49. Queensland Country Life. (1944). Food Front Plan: Queensland Country Life. 6 January, p.2.
50. Rigby Ltd. (1941). Rigby's for Books: LOOK TO THE LAND by Lord Northbourne. Adelaide: The Advertiser. 2 August, p. 10.
51. Rodale, J. I. (1965). Autobiography. Emmaus, MA: Rodale Press.
52. Rodale, J. I. (Ed.). (1942). Organic Farming and Gardening. Emmaus, Pennsylvania: Rodale Press, 1(1), 1-16.
53. Scrutator. (1945). The Future of our Farm Industries. Lismore, NSW: The Northern Star. 7 July, p.4.

54. Selawry, A. (1992). Ehrenfried Pfeiffer: Pioneer of Spiritual Research and Practice. Spring Valley, NY: Mercury Press.
55. SMH. (1940). Land-Love: "Look to the Land" by Lord Northbourne. Sydney: The Sydney Morning Herald (SMH). 21 September, p. 10.
56. Steiner, R. (1924). Agriculture Course ("Printed for private circulation only"; 1929, first English language edition; George Kaufmann Trans ed.). Dornach, Switzerland: Goetheanum.
57. The Executive Officers. (1954). Farewell. Farm & Garden Digest (incorporating Organic Farming Digest), 3(5), 1-3.
58. Till, C. W. (1946). The Economics of Primary Production. Renmark, SA: The Murray Pioneer. 4 July, p.5.
59. West Australian. (1941). Life & Letters. Perth: The West Australian. 22 November, p. 7.
60. White, H. F. (1954). Some aspects of ley farming (An A.B.C. Country Hour broadcast). Farm & Garden Digest (incorporating Organic Farming Digest), 3(5), 3-6.
61. White, H. F. (1959). After 50 Years: Human Life and the Food Chain. Bald Blair, Guyra, NSW. White, H. F. & Hicks, C. S. (1953). Life from the Soil. Melbourne: Longmans Green & Co.

CHAPTER 3

Overview of the Global Spread of Conservation Agriculture

Theodor Friedrich, Rolf Derpsch, and
Amir Kassam

3.1 INTRODUCTION

3.1.1 THE NEED FOR CONSIDERING THE ENVIRONMENTAL FOOTPRINT OF AGRICULTURE

There appears to be no alternative but to increase agricultural productivity (i.e. crop yield per unit area) and the associated total and individual factor productivities (i.e. biological output per unit of total production input, and output per unit of individual factors of production such as energy, nutrients, water, labor, land and capital) to meet the global food, feed, fiber and bio-energy demand and to alleviate hunger and poverty. However, until now, agricultural intensification from intensive tillage-based production systems generally has had a negative effect on the quality of many of the essential natural resources such as soil, water, terrain, biodiversity and the associated ecosystem services provided by nature. This degradation of the land resource base has caused crop yields and factor productivities to decline and has forced farmers, scientists and development stakeholders to search for an alternative paradigm that is ecologically sustainable as well as profitable. Another challenge for agriculture is its environmental footprint and climate change. Agriculture is responsible for about 30% of the total greenhouse gas emissions of CO2, N2O and CH4 while being directly affected by the consequences of a changing climate (IPCC, 2007).

The new paradigm of "sustainable production intensification" as elaborated in FAO (2011a) recognizes the need for a productive and remunerative agriculture which at the same time conserves and enhances the natural resource base and environment, and positively contributes to harnessing the environmental services. Sustainable crop production intensification must not only reduce the impact of climate change on crop production but also mitigate the factors that cause climate change by reducing emissions and by contributing to carbon sequestration in soils. Intensification should also enhance biodiversity in crop production systems above and below the ground to improve ecosystem services for better productivity and healthier environment. A set of soil-crop-nutrient-water-landscape system management practices known as Conservation Agriculture (CA) has the potential to deliver on all of these goals. CA saves on production energy input and mineral nitrogen use in farming and thus reduces emissions; it enhances biological activity in soils, resulting in long-term yield and factor productivity increases. While not tilling the soil is a necessary, but not sufficient condition for truly sustainable and productive agriculture, CA has to be complemented with other techniques, such as integrated pest management, plant nutrient management, and weed and water management (FAO, 2011a).

3.1.2 DEFINITION AND DESCRIPTION OF CONSERVATION AGRICULTURE

According to FAO, Conservation Agriculture (CA) is an approach to managing agro-ecosystems for improved and sustained productivity, increased profits and food security while preserving and enhancing the resource base and the environment. CA is characterized by three linked principles, namely:

- Continuous no or minimal mechanical soil disturbance (i.e., no-tillage and direct sowing or broadcasting of crop seeds, and direct placing of planting material in the soil; minimum soil disturbance from cultivation, harvest operation or farm traffic, in special cases limited strip tillage);
- Permanent organic soil cover, especially by crop residues, crops and cover crops; and

- Diversification of crop species grown in sequence or associations through rotations or, in case of perennial crops, associations of plants, including a balanced mix of legume and non legume crops.

CA principles are universally applicable to all agricultural landscapes and land uses with locally adapted practices. CA enhances biodiversity and natural biological processes above and below the ground surface. Soil interventions such as mechanical tillage are reduced to an absolute minimum or avoided, and external inputs such as agrochemicals and plant nutrients of mineral or organic origin are applied optimally and in ways and quantities that do not interfere with, or disrupt, the biological processes (FAO, 2011b).

CA facilitates good agronomy, such as timely operations, and improves overall land husbandry for rain-fed and irrigated production. Complemented by other known good practices, including the use of quality seeds, and integrated pest, nutrient, weed and water management, etc., CA is a base for sustainable agricultural production intensification. The yield levels of CA systems are comparable with and even higher than those under conventional intensive tillage systems, which means that CA does not lead to yield penalties. At the same time, CA complies with the generally accepted ideas of sustainability. As a result of the increased system diversity and the stimulation of biological processes in the soil and above the surface as well as due to reduced erosion and leaching, the use of chemical fertilizer and pesticides, including herbicides, is reduced in the long term. Ground water resources are replenished through better water infiltration and reduced surface runoff. Water quality is improved due to reduced contamination levels from agrochemicals and soil erosion (Laurent et al., 2011). It further helps to sequester carbon in soil at a rate ranging from about 0.2 to 1.0 t/ha/year depending on the agro-ecological location and management practices (Corsi et al., 2012). Labor requirements are generally reduced by about 50%, which allows farmers to save on time, fuel and machinery costs (Saturnino and Landers, 2002; Baker et al, 2007; Lindwall and Sonntag, 2011; Crabtree, 2010). Fuel savings in the order of around 65% are in general reported (Sorrenson and Montoya, 1984; 1991).

3.1.3 HISTORY, DEVELOPMENT AND RELEVANCE OF CA

Tillage, particularly in fragile ecosystems, was questioned for the first time in the 1930s, when the dustbowls devastated wide areas of the mid-west United States. Concepts for reducing tillage and keeping soil covered came up and the term conservation tillage was introduced to reflect such practices aimed at soil protection. Seeding machinery developments allowed then, in the 1940s, to seed directly without any soil tillage. At the same time theoretical concepts resembling today's CA principles were elaborated by Edward Faulkner in his book "Ploughman's Folly" (Faulkner, 1945) and Masanobu Fukuoka with the "One Straw Revolution" (Fukuoka, 1975). But it was not until the 1960s for no-tillage to enter into farming practice in the USA. In the early 1970s no-tillage farming reached Brazil, where farmers together with scientists transformed the technology into the system which today is called CA. No-tillage and mulching were also tested in the 1970s in West Africa (Greenland, 1975; Lal, 1977, 1976). Yet it took some 20 years before CA reached significant adoption levels in South America and elsewhere. During this time farm equipment and agronomic practices in no-tillage systems were improved and developed to optimize the performance of crops, machinery and field operations. This process is still far from being over as the creativity of farmers and researchers is still producing improvements to the benefits of the system, the soil and the farmer. From the early 1990s CA began to spread exponentially, leading to a revolution initially in the agriculture of southern Brazil, Argentina and Paraguay. During the 1990s this development increasingly attracted attention from other parts of the world, including development and international research organizations such as FAO, CIRAD and some CGIAR centres. Study tours to Brazil for farmers and policy makers, regional workshops, development and research projects were organized in different parts of the world leading to increased levels of awareness and adoption in a number of African countries such as Zambia, Tanzania and Kenya as well as in Asia, particularly in Kazakhstan and China. The improvement of conservation tillage and no-tillage practices within an integrated farming concept such as CA also led to increased adoption, including in industrialized countries, after the end of the millennium, particularly in Canada, Australia, Spain and Finland.

CA crop production systems are experiencing increased interest in most countries around the world. There are only few countries where CA is not practiced by at least some farmers and where there are no local research results available about CA. The total area under CA in 2011 is estimated to be some 125 million hectares (FAO, 2011c). CA is practiced by farmers from the arctic circle (e.g. Finland) over the tropics (e.g. Kenya, Uganda), to about 50° latitude South (e.g. Malvinas/Falkland Islands); from sea level in several countries of the world to 3,000 m altitude (e.g. Bolivia, Colombia), from extremely dry conditions with 250 mm a year (e.g. Morocco, Western Australia), to heavy rainfall areas with 2,000 mm a year (e.g. Brazil) or 3,000 mm a year (e.g. Chile). No-tillage is practiced on all farm sizes from less than half a hectare (e.g. China, Zambia) to thousands of hectares (e.g. Argentina, Brazil, Kazakhstan). It is practiced on soils that vary from 90% sand (e.g. Australia) to 80% clay (e.g. Brazil's Oxisols and Alfisols). Soils with high clay content in Brazil are extremely sticky but this has not been a hindrance to no-till adoption when appropriate equipment is available. Soils which are readily prone to crusting and surface sealing under tillage farming do not present this problem under CA because the mulch cover avoids the formation of crusts. CA has even allowed expansion of agriculture to land areas considered marginal in terms of rainfall or fertility (e.g. Australia, Argentina). All crops can be grown adequately in CA and to the authors' knowledge there has not yet been a crop that would not grow and produce under this system, including root and tuber crops (Derpsch and Friedrich, 2009).

The main barriers to the adoption of CA practices continue to be: knowledge on how to do it (know how), mindset (tradition, prejudice), inadequate policies, for example, commodity based subsidies (EU, US) and direct farm payments (EU), unavailability of appropriate equipment and machines (many countries of the world), and of suitable herbicides to facilitate weed and vegetation management (especially for large scale farms in developing countries) (FAO, 2008; Friedrich and Kassam, 2009).

3.2 GLOBAL AREA AND REGIONAL DISTRIBUTION

Global data of CA adoption are not officially reported, but collected from local farmers' and interest groups. The data are assembled and published by FAO (FAO, 2011c). For the data collection the CA definition is quantified as follows:

- Minimum Soil Disturbance: Minimum soil disturbance refers to low disturbance no-tillage and direct seeding. The disturbed area must be less than 15 cm wide or less than 25% of the cropped area (whichever is lower). There should be no periodic tillage that disturbs a greater area than the aforementioned limits. Strip tillage is allowed if the disturbed area is less than the set limits.
- Organic soil cover: Three categories are distinguished: 30-60%, >60-90% and >90% ground cover, measured immediately after the direct seeding operation. Area with less than 30% cover is not considered as CA.
- Crop rotations / associations: Rotation/association should involve at least 3 different crops. However, repetitive wheat or maize cropping is not an exclusion factor for the purpose of this data collection, but rotation/ association is recorded where practiced.

The worldwide spread of CA in 2011 (about 125 M ha) is shown in Table 3.1, ranking the countries according to area adopted. CA in recent years has become a fast growing production system. While in 1973/74 the system was used only on 2.8 M ha worldwide, the area had grown to 6.2 M ha in 1983/84 and to 38 M ha in 1996/97 [18]. In 1999, worldwide adoption was 45 M ha, and by 2003 the area had grown to 72 M ha. In the last 11 years CA system has expanded at an average rate of around 7 M ha per year from 45 to 125 M ha showing the increased interest of farmers in this production system.

TABLE 3.1 Extent of Adoption of Conservation Agriculture Worldwide (countries with > 100,000 ha)

Country	CA area (ha)
USA	26,500,000
Argentina	25,553,000
Brazil	25,502,000
Australia	17,000,000
Canada	13,481,000
Russia	4,500,000
China	3,100,000
Paraguay	2,400,000
Kazakhstan	1,600,000
Bolivia	706,000
Uruguay	655,100
Spain	650,000
Ukraine	600,000
South Africa	368,000
Venezuela	300,000
France	200,000
Zambia	200,000
Chile	180,000
New Zealand	162,000
Finland	160,000
Mozambique	152,000
United Kingdom	150,000
Zimbabwe	139,300
Colombia	127,000
Others	409,440
Total	**124,794,840**

The growth of the area under CA has been especially significant in South America where the MERCOSUR countries (Argentina, Brazil, Paraguay and Uruguay) are using the system on about 70% of the total cultivated area. More than two thirds of no-tillage practiced in MERCOSUR is permanently under this system, in other words once started the soil is never tilled again.

As Table 3.2 shows 45% of the total global area under CA is in South America, 32% in the United States of America and Canada, 14% in Australia and New Zealand and 9% in the rest of the world including Europe, Asia and Africa. The latter are the developing continents in terms of CA adoption. Despite good and long lasting research in these continents showing positive results for no-tillage systems, CA has experienced only small rates of adoption.

TABLE 3.2 Area under CA by continent

Continent	Area (ha)	Percent of total
South America	55,464,100	45
North America	39,981,000	32
Australia & New Zealand	17,162,000	14
Asia	4,723,000	4
Russia & Ukraine	5,100,000	3
Europe	1,351,900	1
Africa	1,012,840	1
World total	**124,794,840**	**100**

Because of the benefits that CA systems generate in terms of yield, sustainability of land use, incomes, timeliness of cropping practices, ease of farming and ecosystem services, the area under CA systems has been growing exponentially, largely as a result of the initiative of farmers and their organizations.

Except in a few countries (USA, Canada, Australia, Brazil, Argentina, Paraguay, and Uruguay), however, CA has not been "mainstreamed" in agricultural development programs or backed by suitable policies and institutional support. Consequently, the total arable area under CA worldwide is still relatively small (about 9%) compared to areas farmed using tillage. Nonetheless, the rate of adoption globally since 1990 has been growing exponentially, mainly in North and South America and in Australia and New Zealand. The area under

CA is on the increase in all parts of Asia, and large areas of agricultural land are expected to switch to CA in the coming decade as is already occurring in China, Kazakhstan, and most likely in India.

Although much of the CA development to date has been associated with rain-fed arable crops, farmers can apply the same principles to increase the sustainability of irrigated systems, including those in semi-arid areas. CA systems have also been tailored for orchard and vine crops with the direct sowing of field crops, cover crops and pastures beneath or between rows, giving permanent cover and improved water infiltration, soil aeration and biodiversity. The common constraint mentioned by farmers to practicing this latter type of intercropping is competition for soil water between trees and crops. However, careful selection of deep rooting tree species and shallow rooting annuals resolves this. Also, as there is less runoff, more water enters the soil thereby improving water use efficiency. Functional CA systems do not replace current good land husbandry practices but integrate with them instead.

CA can be seen as an alternate approach to ecologically underpin production systems to enhance productivity, sustainability and resilience. However, introduction and adoption of CA must overcome a range of constraints that have been highlighted by a number of stakeholders (e.g., FAO, 2008).

3.2.1 ADOPTION IN THE AMERICAS

17CA adoption is highest in the southern parts of South America and in the North-Western Parts of North America with adoption levels above 50%. In Canada, with 13.5 M ha under CA, long term and wide adoption of Conservation Agriculture has resulted in visible environmental benefits, including the disappearance of dust storms as well as a higher biodiversity (Lindwall and Sonntag, 2010). Environmental services provided through CA are increasingly recognized, for example through carbon payment schemes as in Alberta, Canada. In the USA CA adoption of 26.5 M ha is still at a significantly lower level in percent of the cropland (16 %), despite experience with no-till for a long period of time. However, for a number of reasons, including commodity-focused subsidies, permanent no-till is applied only on about 10 to 12% of the area under no-tillage. Yet, in the USA the awareness about crop rotations and cover crops as well as the additional benefits of permanent no-till systems is also growing as a result of organized farmers' associations such as the Conservation Agriculture Systems Alliance (CASA).

In Latin America the adoption levels of no-till farming in Argentina, Paraguay, Uruguay and Southern Brazil are approaching the 100 % mark (Table 3). However, there are serious concerns about the quality of the CA adoption. Following market pressures, which are partly increased by government policies, a considerable proportion of farmers is opting for soya monocropping, even without any cover crops between two soya crops, which, despite applying no-till, results in erosion and soil degradation and cannot be considered as "real" CA. Taking this situation into account, the area under good quality CA is less than half of the total area under no-till cropping, particularly in Argentina and Uruguay. The problem is being addressed in Brazil and Uruguay with strengthened extension, legal regulations for cover crops in the specific case of soya and subsidy programs for good quality CA.

TABLE 3.3 CA adoption in some selected countries of Latin America

Country	CA area (ha)
Argentina	25,553,000
Bolivia	706,000
Brazil	25,502,000
Chile	180,000
Colombia	127,000
Mexico	41,000
Paraguay	2,400,000
Uruguay	655,100
Venezuela	300,000
Total	**55,464,100**

3.2.2 ADOPTION IN EUROPE

CA is not widely spread in Europe, excluding Russia (Table 3.4): no-till systems do not exceed 1% of the arable cropland. Only Africa has a smaller absolute area under CA than Europe. Since 1999 ECAF (European Conservation Agriculture Federation) has been promoting CA in Europe, and adoption is visible in Spain, Finland, France and UK, with some farmers at 'proof of concept' stage in Ireland,

Portugal, Germany, Switzerland, and Italy. Especially in Spain, Portugal and Italy the growth of CA in perennial crops, such as fruit orchards, vineyards and olive plantations, has exceeded the adoption rate in annual crops.

TABLE 3.4 CA adoption in some selected countries of Europe

Country	CA area (ha)
Finland	160,000
France	200,000
Germany	5,000
Hungary	8,000
Ireland	100
Italy	80,000
Netherlands	500
Portugal	32,000
Slovakia	10,000
Spain	650,000
Swilzerland	16,300
United Kingdom	150,000
Total	**1,311,900**

Bridging between Europe and Asia are two countries, Russia and Ukraine, with significant adoption of CA and with active farmers' groups promoting CA. In Russia the area under conservation tillage is reported with 15 M ha, while CA according to the FAO definition is applied on 4.5 M ha. In the Ukraine CA has reached 600,000 ha.

3.2.3 ADOPTION IN ASIA

Asian countries have seen increasing uptake of CA in the past 10-15 years. In Central Asia, a fast development of CA has been observed in the last 5 years in Kazakhstan which now has 10.5 M ha under reduced tillage, mostly in the northern drier provinces, and of this 1.6 M ha (10% of crop area) are "real" CA with permanent no-till and rotation that puts Kazakhstan amongst the top ten countries in the world with the largest crop area under CA systems. Chinatoo has had an equally dynamic development of CA which began 20 years ago with research and then increased adoption during the last few years, including extending the CA system to rice production. Now more than 3.1 M ha are under CA in China and 23,000 ha in DPR Korea where the introduction of CA has made it possible to grow two successive crops (rice or maize or soya as summer crop, winter wheat or spring barley as winter crop) within the same year, through direct drilling of the second crop into the stubble of the first (Table 3.5).

TABLE 3.5 CA adoption in some selected countries of Asia

Country	CA area (ha)
China	3,100,000
Kazakhstan	1,600,000
Korea, DPR	23,000
Total	**4,723,000**

In the Indo-Gangetic Plains across India, Pakistan, Nepal and Bangladesh, in the wheat-rice cropping system, there is large adoption of no-till wheat with some 5 M ha, but only marginal adoption of permanent no-till systems and full CA. In India, the adoption of no-till practices by farmers has occurred mainly in the wheat portion of the wheat-rice double cropping system.

3.2.4 ADOPTION IN WEST ASIA AND NORTH AFRICA

In the WANA (West Asia and North Africa) region, much of the CA work done in various countries has shown that yields and factor productivities can be improved with no-till systems. Extensive research and development work has been conducted in several countries in the region since the early 1980s such as in Morocco, Tunisia, Syria, Lebanon and Jordan, and in Turkey (Table 3.6). While Morocco and particularly Tunisia showed a modest growth in CA adoption, the uptake has literally exploded in Syria, spreading over nearly 20,000 ha in only few years. The main reason for the rapid uptake has been the shortage of fuel and increased availability of locally produced affordable no-till seeders, which are now being exported to other countries in the region, and the efforts of development and promotion activities by organization such as GIZ, ICARDA, ACSAD and Aga Khan Foundation.

Key lessons from international experiences about CA and considerations for its implementation in the Mediterranean region show the potential benefits that can be harnessed by farmers in the semi-arid Mediterranean environments while highlighting the need for longer-term research including on weed management, crop nutrition and economics of CA systems. In addition, it is clear that without farmer engagement and appropriate enabling policy and institutional support to achieve effective farmer engagement and a process for testing CA practices and learning how to integrate them into production system, rapid uptake of CA is not likely to occur.

TABLE 3.6 CA adoption in selected countries of North Africa and Near East

Country	CA area (ha)
Lebanon	1,200
Morocco	4,000
Syria	18,000
Tunisia	8,000
Total	**31,200**

Work by ICARDA and CIMMYT has shown benefits of CA especially in terms of increase in crop yields, soil organic matter, water use efficiency and net revenue. CA also shows the importance of utilizing cropping and crop diversification with legumes and cover crops instead of a fallow period, providing improved productivity, soil quality, N-fertilizer use efficiency and water use efficiency. CA is perceived as a powerful tool of land management in dry areas. It allows farmers to improve their productivity and profitability especially in dry areas while conserving and even improving the natural resource base and the environment. However, CA adaptation in drylands faces critical challenges linked to water scarcity and drought hazard, low biomass production and acute competition between conflicting uses including for soil cover, animal fodder, cooking/heating fuel, raw material for habitat etc. Poverty and vulnerability of many smallholders that rely more on livestock than on grain production are other key factors.

3.2.5 ADOPTION IN SUB-SAHARAN AFRICA

In the Sub-Saharan Africa, innovative participatory approaches are being used to develop supply-chains for producing CA equipment targeted at small holders.

Similarly, participatory learning approaches such as those based on the principles of farmer field schools (FFS) are being encouraged to strengthen farmers' understanding of the principles underlying CA and how these can be adapted to local situations.

CA is now beginning to spread to Sub-Saharan Africa region, particularly in eastern and southern Africa as can be seen in Table 3.7. Building on indigenous and scientific knowledge and equipment design from Latin America, and, more recently, with collaboration from China, Bangladesh and Australia, farmers in at least 14 African countries are now using CA (in Kenya, Uganda, Tanzania, Sudan, Swaziland, Lesotho, Malawi, Madagascar, Mozambique, South Africa, Zambia, Zimbabwe, Ghana and Burkina Faso). CA has also been incorporated into the regional agricultural policies by NEPAD (New Partnership for Africa's Development).

TABLE 3.7 CA adoption in Sub-Saharan Africa

Country	CA area (ha)
Ghana	30,000
Kenya	33,000
Lesotho	2,000
Malawi	16,000
Madagascar	6,000
Mozambique	152,000
Namibia	340
South Africa	368,000
Sudan	10,000
Tanzania	25,000
Zambia	200,000
Zimbabwe	139,300
Total	**981,640**

In the specific context of Africa with resource-poor farmers, CA systems are relevant for addressing the challenges of climate change, high energy costs, environmental degradation, and labor shortages. So far the area under CA is small, but there is a steadily growing movement that already involves more than 400,000 small-scale farmers in the region for a total area of nearly 1 M ha.

In Sub-Saharan Africa CA is expected to increase food production while reducing negative effects on the environment and energy costs, and to result in the development of locally-adapted technologies consistent with CA principles.

3.3 CONCLUSIONS

CA represents the core components of a new alternative paradigm for the 21st century and calls for a fundamental change in production system thinking. It is counterintuitive, novel and knowledge and management intensive. The roots of the origins of CA lie more in the farming communities than in the scientific community, and its spread has been largely farmer-driven. Experience and empirical evidence across many countries has shown that the rapid adoption and spread of CA requires a change in commitment and behavior of all concerned stakeholders. For the farmers, a mechanism to experiment, learn and adapt is a prerequisite. For policy-makers and institutional leaders, transformation of tillage systems to CA systems requires that they fully understand the large and longer-term economic, social and environmental benefits CA paradigm offers to the producers and the society at large. Further, the transformation calls for a sustained policy and institutional support role that can provide incentives and required services to farmers to adopt CA practices and improve them over time (FAO, 2008; Friedrich and Kassam, 2009; Friedrich et al., 2009; Kassam et al., 2009, 2010). Originally the adoption of CA was mainly driven by acute problems faced by farmers, especially wind and water erosion, as for example southern Brazil or the Prairies in North America, or drought as in Australia. In all these cases farmers' organization was the main instrument to generate and spread knowledge that eventually led to mobilizing public, private and civil sector support. More recently, again pressed by erosion and drought problems, exacerbated by increase in cost of energy and production inputs, government support has played an important role in accelerating the adoption rate of CA, leading to the relatively fast adoption rates for example in Kazakhstan and China, but also in African countries such as Zambia and Zimbabwe, among others, and this is attracting support from other stakeholders.

Today the main reasons for adoption of CA can be summarized as follows: (1) better farm economy (reduction of costs in machinery and fuel and time-saving in the operations that permit the development of other agricultural and non-agricultural complementary activities); (2) flexible technical possibilities for sowing, fertilizer application and weed control (allows for more timely operations); (3) yield increases and greater yield stability (as long term effect); (4) soil protection against water and wind erosion; (5) greater nutrient-efficiency; and (6) better water economy in dryland areas. Also, no-till and cover crops are used between rows of perennial crops such as olives, nuts and grapes. CA can be used for winter crops, and for traditional rotations with legumes, sunflower and canola, and in field crops under irrigation where CA can help optimize irrigation system management to conserve water, energy and soil quality, reduce salinity problems and to increase fertilizer use efficiency.

At the landscape level, CA enables several environmental services to be harnessed at a larger scale, particularly C sequestration, cleaner water resources, drastically reduced erosion and runoff, and enhanced biodiversity. Overall, CA as an alternative paradigm for sustainable production intensification offers a number of benefits to the producers, the society and the environment that are not possible to obtain with tillage agriculture (Kassam et al., 2010). So, CA is not only climate-smart, it's smart in many other ways.

BIBLIOGRAPHY

1. Baig, M.N., Gamache, P.M. (2009), The Economic, Agronomic and Environmental Impact of No-Till on the Canadian Prairies, Alberta Reduced Tillage Linkages, Canada.
2. Baker, C.J., Saxton, K.E., Ritchie, W.R., Chamen, W.C.T., Reicosky, D.C., Ribeiro, M.F.S., Justice, S.E., Hobbs, P.R. (2007), No-Tillage Seeding in Conservation Agriculture – 2nd edn. CABI and FAO. 326 pp.
3. Corsi, S., Friedrich, T., Kassam, A., Pisante, M., de Moraes Sà, J. (2012), Soil organic carbon accumulation and greenhouse gas emission reductions from Conservation Agriculture: A literature review. Integrated Crop Management Vol. 16, FAO, Rome, Italy. 89 pp.
4. Crabtree, B. (2010), In Search for Sustainability in Dryland Agriculture, Crabtree Agricultural Consulting, Australia. 204 pp.
5. Derpsch, R. (1998), Historical review of no-tillage cultivation of crops, in: Proceedings of the 1st JIRCAS Seminar on Soybean Research on No-tillage Culture & Future Research Needs, JIRCAS Working Report No. 13, pp. 1-18, Iguassu Falls, Brazil, March 5-6, 1998.
6. Derpsch, R., Friedrich, T. (2009), Development and Current Status of No-till Adoption in the World, Proceedings on CD, 18th Triennial Conference of the International Soil Tillage Research Organization (ISTRO), Izmir, Turkey, June 15-19, 2009.
7. FAO (2008), Investing in Sustainable Crop Intensification: The Case for Soil Health. Report of the International Technical Workshop, FAO, Rome, July. Integrated Crop Management, Vol. 6. Rome: FAO. Online at: http://www.fao.org/ag/ca/.
8. FAO (2011a), Save and Grow, a policymaker's guide to sustainable intensification of smallholder crop production, Food and Agriculture Organization of the United Nations, Rome. 116 pp.

9. FAO (2011b), What is Conservation Agricutlure? FAO CA website (http://www.fao.org/ag/ca/1a.html), FAO, Rome.
10. FAO (2011c), CA Adoption Worldwide, FAO-CA website available online at: (http://www.fao.org/ag/ca/6c.html).
11. Faulkner, E.H. (1945), Ploughman's Folly, Michael Joseph, London. 142 pp.
12. Fukuoka, M. (1975), One Straw Revolution, Rodale Press, English translation of shizen noho wara ippeon no kakumei, Hakujusha Co., Tokyo. 138 pp.
13. Greenland, D. J. (1975), Bringing the green revolution to the shifting cultivators. Science 190: 841-844. DOI : 10.1126/science.190.4217.841
14. IPCC (2007), Climate Change; Fourth Assessment report of the Intergovernmental Panel on Climate Change, Cambridge University Press.
15. Friedrich, T., Kassam, A.H. (2009), Adoption of Conservation Agriculture Technologies: Constraints and Opportunities, Proceedings of the IV World Congress on Conservation Agriculture, ICAR, New Delhi, India, 4-7 February, 2009.
16. Friedrich, T., Kassam, A.H., Shaxson, F. (2009), Conservation Agriculture, in: Agriculture for Developing Countries, Science and Technology Options Assessment (STOA) Project, European Technology Assessment Group, Karlsruhe, Germany.
17. Kassam, A.H., Friedrich, T., Shaxson, F., Pretty, J. (2009), The spread of Conservation Agriculture: justification, sustainability and uptake, International Journal of Agricultural Sustainability 7(4):1-29. DOI : 10.3763/ijas.2009.0477
18. Kassam, A.H., Friedrich, T., Derpsch, R. (2010), Conservation Agriculture in the 21st century: A paradigm of sustainable agriculture, Proceedings of the European Congress on Conservation Agriculture, Madrid, Spain, October 2010.
19. Lal, R.(1975), Role of mulching techniques in tropical soil and water management. IITA Technical Bulletin 1, Ibadan, Nigeria, 38 pp.
20. Lal, R (1976), No tillage effects on soil properties under different crops in western Nigeria. Soil Sci. Soc. Amer. Proc. 40: 762-768.
21. Laurent, F., Leturcq, G., Mello, I., Corbonnois, J., Verdum, R. (2011), La diffusion du semis direct au Brésil, diversité des pratiques et logiques territoriales: l'exemple de la région d'Itaipu au Paraná. Confins 12 [online]. URL : http://confins.revues.org/7143. DOI: 10.4000/confins.7143
22. Lindwall, C.W., Sonntag, B. (eds.) (2010), Landscape Transformed: The History of Conservation Tillage and Direct Seeding, Knowledge Impact in Society, Saskatoon, University of Saskatchewan.
23. Saturnino, H.M., Landers, J.N. (2002), The Environment and Zero Tillage, APDC-FAO, Brasilia, Brazil, UDC. 139 pp.
24. Sorrenson, W.J., Montoya, L.J. (1984), Implicaçoes econômicas da erosao do solo e de practices conservacionistas no Paraná, Brasil, IAPAR, Londrina, GTZ, Eschborn. 231 pp.
25. Sorrenson, W.J., Montoya, L.J. (1991), The economics of tillage practices, in: R. Derpsch, C. H. Roth, N. Sidiras, U. Kopke, Controle da erosão no Paraná, Brasil: sistemas de cobertura do solo, plantio direto e preparo conservacionista do solo, GTZ, Eschborn, pp. 165 -192.

CHAPTER 4

The Transition from Green to Evergreen Revolution

M. S. Swaminathan and P. C. Kesavan

4.1 INTRODUCTION

The Green Revolution of the 1960s would remain in the annals of the globe as the most positive achievement of humankind after the destructive world wars, the Holocaust, detonation of atomic bombs over Hiroshima and Nagasaki and the great Bengal famine of 1943-1944. To India, in particular, the backdrop of the Bengal famine did not augur well when it attained independence in August 1947. This led to the first Prime Minister of India Jawaharlal Nehru, declaring that 'everything else can wait, but not agriculture.' The task ahead of the agricultural scientists was immense, because the goal was to enhance the productivity in terms of kg per hectare rather than the total production by bringing more of forestland under cultivation. Increasing the area under irrigation was yet another approach to achieve more total production, and this also was done. There is certainly a limit to increasing land under cultivation and tapping all sources of irrigation. Further, the Indian agriculture is a 'gamble in the monsoon'. The ideal approach was to increase the productivity (kg/ha) of the crops by making them more responsive to external inputs of inorganic mineral fertilizers commonly referred to as N, P, K (viz. nitrogen, phosphorus and potassium respectively). Inorganic nitrogen fertilizer was particularly important. Yet, it was not simple to saturate the soil with inorganic nitrogen fertilizers as the Indian varieties of wheat and rice (i.e. the two staple food crops) with their characteristic tall stalks (stem) and dense and long panicles lodged (i.e. fell back) under the increased weight of the grains. So, an appropriate plant type would be dwarf/semi-dwarf plants with panicles of normal length.

Early Indian work in the 1950s at the Central Rice Research Institute (CRRI), Cuttack, India, involved making crosses between short 'japonica' rice varieties and the tall indica rice varieties. The goal was to make the tall Indian rice varieties shorter and more robust to withstand the weight of the heavy panicles of plants grown in soils enriched with about 100 kg N2 per hectare. This led to the development of a few varieties of rice like Mashuri and ADT-27 more responsive than pure indica rices to exogenous inputs of inorganic nitrogen fertilizers. Hence, it is reasonable to view that seeds of 'Green Revolution' were first sown in the fields of CRRI, Cuttack, India. What then followed in the wheat fields of the Indian Agricultural Research Institute, New Delhi, became remarkable for at least two reasons. One is that dwarf and semi-dwarf wheats with dwarfing genes from Norin-10 wheat (Japan) were highly fertilizer-responsive and they broke the 'yield-ceiling' of Indian wheat varieties. Consequently the then image of India as 'begging bowl' changed to 'bread basket'. The second reason is that the Green Revolution silenced the doomsayers who wrote that the famine of food and consequent hunger in India would never be solved and that a vast section of humanity would perish in hunger. In this regard, references are made to the two books, *The Population Bomb*, by Ehrlich [1] and *Famine 1975: America's Decision: Who Will Survive?* by William and Paul Paddock [2]. A serious question in the mind of one of us (MSS) who was deeply associated with late Norman Borlaug in ushering in the India's Green Revolution was whether the Green Revolution type of agriculture, which indeed was 'exploitative agriculture' could be sustainable over long periods of time. Convinced that it would not be sustainable, and should only be used to get some 'breathing space', Swaminathan [3] emphasized this point in his Chairman's Address to the Agricultural Section of the 55th Indian Science Congress Session held in January 1968 in Varanasi, India. What he then said is reproduced below, since it is still valid.

Intensive cultivation of land without conservation of soil fertility and soil structure would lead ultimately to the springing up of deserts. Irrigation without arrangements for drainage would result in soils getting alkaline or saline. Indiscriminate use of pesticides, fungicides and herbicides could cause adverse changes in biological balance as well as lead to an increase in the incidence of cancer and other diseases, through the toxic residues present in the grains or other edible parts. Unscientific tapping of underground water would lead to the rapid exhaustion of this wonderful capital resource left to us through ages of natural farming. The rapid replacement of numerous locally adapted varieties with one or two high-yielding strains in large contiguous areas would result in the spread of serious diseases capable of wiping out entire crops, as happened prior to the Irish potato famine of 1845 and the Bengal rice famine of 1943. Therefore, the initiation of exploitative agriculture without a proper understanding of the various consequences of every one of the changes introduced into traditional agriculture and without first building up a proper scientific and training base to sustain it, may only lead us into an era of agricultural disaster in the long run, rather than to an era of agricultural prosperity [3].

Further, what was described as exploitative agriculture [3] was, however, referred to as 'Green Revolution' by William Gaud of the US Agency for International Development. More than four decades later, Bourne [4] wrote, "The Green Revolution Borlaug started had nothing to do with the eco-friendly green label in vogue today". In the following year, Dhillon., et al. [5] demonstrated with the help of a wealth of data that the productivity of wheat and rice in Punjab had been clearly plateauing since 1996-1997 through 2007-2008. Yet another disconcerting aspect of the Green Revolution was that it did not ensure food security at the household level to hundreds of millions of resource-poor small and marginal farming, fishing and landless families. The paradox, "mountains of grains on one hand and millions of hungry people on the other" aptly described the factual position. The Green Revolution provided only enhanced availability of food at the national level, but was not designed to fight the 'famine of rural livelihoods' in order to enhance 'access' (i.e. purchasing power) to food. Attention was also not given to the provision of clean drinking water, so essential to avoid gastric diseases and retain the ingested food for 'absorption' or utilization in the body. These are the primary considerations which led to MSS to develop a 'systems approach'–based 'evergreen revolution' in order to achieve productivity in perpetuity without ecological and social harm. The systems approach involves concurrent attention to all the ecological foundations of sustainable agriculture, such as soil structure and soil health, freshwater, biodiversity, renewable energy and climate. The basic principles, design and goals of the 'evergreen revolution' are described in several publications by Swaminathan [6-11]. Kesavan and Swaminathan [12-16] have elaborated as to how the evergreen revolution would be able to fight both the 'famines of food and rural livelihoods' in an ecofriendly and socially 'inclusive' manner. They have also shown [14] that evergreen revolution which has all the elements of sustainable rural development could be integrated with disaster preparedness to reduce loss of lives and livelihoods following a major natural disaster. With particular reference to hydro-meteorological extreme events (e.g. cyclones, floods, droughts etc), sustainable rural development greatly enhances the resilience or the coping capacity of the vulnerable rural communities.

4.2 STRUCTURE AND OUTCOMES OF THE EVERGREEN REVOLUTION

'Evergreen revolution' does not deal only with cultivation of crops as does the Green Revolution. In other words, it is not commodity-centric. Besides various forms of ecoagriculture [12], the evergreen revolution includes sustainable management of natural resources and development of on-farm and non-farm livelihoods using ecotechnologies within a 'biovillage' paradigm and knowledge empowerment of the rural people through modern information and communication technology (ICT) based Village Knowledge Centres [13,15,16].

With regard to ecoagriculture, the emphasis is on 'organic agriculture' and 'green agriculture'. The organic agriculture as defined by the International Federation of Organic Agriculture Movements (IFOAM) relies on ecological processes, biodiversity and cycles adapted to local conditions, rather than the use of chemical inputs with adverse effects. A more flexible system is the Green Agriculture which permits the use of chemical inputs within the schedules of integrated pest management (IPM), integrated nutrient management (INM) and natural resource management systems.

Various modifications of the organic agriculture, such as 'Effective Microorganisms (EM) agriculture', 'one-straw revolution' (system of natural farming without ploughing, chemical fertilizers, weeding and chemical pesticides and herbicides) and 'white agriculture' (system of agriculture based on substantial use of microorganisms, particularly fungi) have been described by Kesavan and Swaminathan [12,15]. None of these involves the use of chemical fertilizers and chemical pesticides.

The evergreen revolution, most importantly, integrates farm animals with eco-fisheries/eco-aquaculture (in the coastal villages) for dung, urine, draught power, milk and meat. Dung is used to produce methane by anaerobic fermentation (pyrolysis) and the methane is used as cooking gas. Used in this manner, methane is not released as a potent greenhouse gas, and furthermore, with the provision of methane as a cooking gas, the rural women do not have to fell the trees as fuel wood or walk long distances to gather biomass for fuel. Emission of greenhouse gases (GHGs) due to cutting down of trees is also arrested, a step in the direction of mitigation.

Recommended for resource-poor small and marginal farms, the evergreen revolution provides multiple sources of income from cereal and cash crops, eggs, milk and meat and also provides renewable energy. The year 2014 was declared the 'International Year of Family Farming'. Writing an editorial in Science on 'Zero hunger' Swaminathan [17] suggested that family farming characterized by diversified crops can be harnessed to support nutrition sensitive agriculture. Just following this, Kesavan and Swaminathan [18] elaborated as to how the evergreen revolution with several elements of family farming could cultivate crops identified for their specific nutrient content in order to provide agricultural remedies to nutritional maladies in different agro-ecological regions. Within the framework of evergreen revolution, a farming system to leverage agriculture for nutritional outcomes, called "farming system for nutrition" (Swaminathan and his co-workers [19] would help to integrate food security (i.e. under nutrition due to inadequate intake of food) with nutrition security (i.e. addressing the hidden hunger caused by the deficiency of vitamin A, vitamin B, zinc, iron, iodine etc. The other unique features of the evergreen revolution, not envisaged in the Green Revolution, is the skill and knowledge empowerment of hundreds of millions of resource-poor small and marginal farmers. The purpose is to enable them to create and manage on-farm and non-farm rural livelihoods to generate income and thus, to enhance 'access' to food. The skill empowerment is imparted to the rural women and men through a pedagogic method of 'learning by doing' that was called 'techniracy' by Swaminathan [20]. The technologies used in the rural areas both for sustainable management of natural resources and creation of on-farm and non-farm livelihoods are called the 'ecotechnologies' which are the resultant of blending frontier technologies (e.g. space, modern information and communication, nuclear, nano-, modern biotechnology etc) with traditional knowledge and ecological prudence of the rural and tribal people, especially the women. The ecotechnologies have a pro-nature, pro-poor, pro-women and pro-employment orientation. What this means is that technologies are designed to have ecological, social as well as gender orientations. It is known that subsistence farming over centuries in thousands of villages in India has not been adequately productive to leave marketable surplus beyond the household requirement of food grains, pulses, oilseeds etc. The bulk of India's poor are among these subsistence farmers. And about 60 percent of India's population lives in about 638,000 villages. In the early design of the evergreen revolution, and the setting up of the M.S. Swaminathan Research Foundation in 1988, the Founder-Chairman (M.S.

Swaminathan) focused on revitalizing the largely indolent villages with subsistence farming into vibrant agri-business-oriented villages. So, he came up with the concept of 'biovillages' (bios = living) which are the centres where 'ecotechnologies' are harnessed for sustainable management of locally-available natural resources, and also to develop one or more ecotechnology driven on-farm and non-farm ecoenterprises with market linkages for income generation. Groups of 10-15 women only or women plus men organize themselves into self-help groups (SHGs) who are then given training, capacity and initial requirement of resources etc., to adopt one or more of several ecoenterprises. The SHGs also require the power of scale to succeed in their ventures. In the globalized trade, the policies and preferences are in favour of 'mass production' than 'production by masses'. However, India's milk production is an example of how 'production by masses' has made India the world leader in the production of milk in the world (138 million tons in 2014 -15) and provides directly and indirectly livelihoods for about 18 million rural people of which women constitute nearly 70 percent. On the other hand, factory/corporate industrial farming in milk production would have displaced about 50-60 percent of the rural women and men from dairy-related livelihoods. USA is a good example.

Knowledge empowerment is quite essential for sustainable and productive living today. Particularly, evergreen agriculture requires the support of biological software for marrying productivity and sustainability. The biological software includes bio-pesticides, bio-fertilizers, vermiculture etc. Hundreds of millions of resource-poor small and marginal farming, fishing and landless rural women and men are often in desperate need of relevant information to save their crops from an impending disease or pest attack, or their cows from a complicated, life-threatening delivery, market prices for their local produce, hospitals for emergency and scores of routine daily needs. In the case of marginal fishers venturing on sea in their country rafts, information on weather, sea wave heights, and fish shoals is very important both for successful operation and personal safety. So, modern information and communication technology-based village knowledge centres (VKCs) were set up in several villages. Young women who have passed the 8th or 9th class are given training in computer operation and use of different programmes and softwares. They become very adept in using Internet for linking data seekers (i.e. rural people) with data holders (i.e. scientific institutions, state and central government establishments, media, non-governmental agencies). The VKCs designed by the MSSRF and managed by the local grass root institutions promote lab-to-lab, lab-to-land, land-to-lab and land-to-land interactions. Special attention is given to update information content on weather, crop and animal husbandry, market trends and prices, welfare schemes, education, employment, healthcare etc. The mobile phones have greatly facilitated the provision of information to far greater number of people instantaneously.

The ecoagriculture/family farming system which is an integral part of the evergreen revolution movement does not necessarily cultivate only the improved high-yielding varieties; instead, the bulk of the cultivation involves locally-adapted land races and indigenous varieties. Hence, it contributes to the conservation of the precious genes and gene pools on the one hand, and consumption and commerce on the other. Swaminathan [21] has proposed a continuum of conservation, cultivation, consumption and commerce (the 4Cs) as the pathway to link biodiversity conservation with food and nutrient security at the household level towards the goal of Biohappiness.

The evergreen revolution that encourages the cultivation of locally-adapted cultivars is innately more resilient to a variety of biotic and abiotic stresses induced by global warming induced climate change. Varied effects of climate change on the growth and productivity of crop plants have been discussed by Swaminathan and Kesavan [15] in detail elsewhere. Impact of climate change on agriculture has been discussed in detail elsewhere by Swaminathan and Kesavan [22].

4.3 CONCLUSIONS

The 'Green Revolution' that changed India's image as 'begging bowl' to 'bread basket' was commodity-centric, and it was not integrated with ecological and social dimensions of sustainable agriculture. Consequently, the productivity has been plateauing since the mid-1990s.

The evergreen revolution, on the other hand, is based on a 'systems approach' with concurrent attention to all the ecological foundations (i.e. soil, freshwater, renewable energy, biodiversity, climate etc) of agriculture and socio-economic dimension of access to food and nutrition security.

With the principles of ecoagriculture, ecotechnologies based ecoenterprises of on-farm and non-farm livelihoods and renewable energy sources, the evergreen revolution is ideal for transforming the subsistence farming into vibrant agri-business enterprises.

As with family farming, the ecoagriculture of evergreen revolution is conducive to integrate food security with nutrition security in the nature of 'farming system for nutrition' (FSN) and provide agri-horticultural remedies to nutritional maladies in the different agro-climatic zones.

With its continuum of conservation, cultivation, consumption and commerce (4Cs), the evergreen revolution is the pathway to 'biohappiness', i.e. happiness arising from the sustainable and equitable conversion of biodiversity into jobs and incomes. Ever-green agriculture promotes the growth of biological software industries, which are environment-friendly and lend themselves to the "production by masses" pathway of job-led economic growth.

BIBLIOGRAPHY

1. Ehrlich P. "The Population Bomb". Ballantine Books, New York (1968): 201.
2. Paddock W and Paddock P. "Famine 1975: America's Decision: Who will survive?" Little Brown and Co. (1967): 286.
3. Swaminathan MS. "The Age of Agony, Genetic Destruction of Yield Barriers and Agricultural Transformation Presidential Address Section of Agricultural Sciences". Proc. 55th Indian Science Congress Varanasi, India (1968): 236-248.
4. Bourne JK. "The Global Food Crisis The End of Plenty". National Geographic Magazine (2009):
5. Dhillon BS., et al. "National Food Security vis-à-vis sustainability of agriculture in high crop productivity regions". Current Science 98 (2010): 33-36.
6. Swaminathan MS. "Sustainable Agriculture: Towards an Evergreen Revolution". Konark Publishers Pvt. Ltd. Delhi, (1996): 232.
7. Swaminathan MS. "Sustainable Agriculture: Towards Food Security". Konark Publishers Pvt. Ltd. Delhi (1996): 272.
8. Swaminathan MS. "I predict: A Century of Hope: Towards an Era of Harmony with Nature and Freedom from Hunger". East West Books (Madras) Pvt. Ltd. (1999): 155.
9. Swaminathan MS. "An Evergreen Revolution". Biologist 47 (2000): 85-89.
10. Swaminathan MS. "Science and Sustainable Food Security - Selected papers of M.S. Swaminathan". Indian Institute of Science (IISc) Centenary Lecture Series 3 (2010): IISc Press. World Scientific, New Jersey, London, Hongkong, Chennai, (2010): 420.
11. Swaminathan MS. "From Green to Evergreen Revolution. Indian Agriculture: Performance and Challenges". Academic Foundation, New Delhi (2010): 410.
12. Kesavan PC and Swaminathan MS. "From Green Revolution to Evergreen Revolution: Pathways and terminologies". Current Science 90 (2006): 145-146.
13. Kesavan PC and Swaminathan MS. "Strategies and models for agricultural sustainability in developing Asian countries". Philosophical Transactions of the Royal Society B: Biological (2008): 877-891.
14. Kesavan PC and Swaminathan MS. "Managing extreme natural disasters in coastal areas". PhilosophicalTransactions of the Royal Society A: Biological 364 (2006): 2191-2216.
15. Kesavan PC and Swaminathan MS. "Evergreen Revolution in Agriculture: Pathway to green economy". Westville Publishing House, New Delhi (2012): 139.

16. Kesavan PC and Swaminathan MS. "Future Direction of Asian Agriculture: Sustaining Productivity without Ecological Degradation". ASM Science Journal 1.2 (2007): 161-168.
17. Swaminathan MS. "Zero Hunger" Science 345 (2014): 461.
18. Kesavan PC and Swaminathan MS. "International Year of Family Farming: a boost to evergreen revolution". Current Science107 (2014): 1970-1974.
19. Das PK., et al. "A Farming System Model to Leverage Agriculture for Nutritional outcomes". AgricultureResearch 3.3 (2014): 193-203.
20. Swaminathan MS. "Agricultural Evolution, Productive Employment and Rural Prosperity". The Princess Leelavathi Memorial Lecture: University of Mysore (1972): 35.
21. Swaminathan MS. "In search of Biohappiness: Biodiversity and Food, Health and Livelihood Security". 2nd edition World Scientific Publishing Co. Pvt. Ltd. Singapore (2015): 184.
22. Swaminathan MS and Kesavan PC. "Agricultural Research in an era of climate change". Agriculture Research 1.1 (2012): 3-11.

CHAPTER 5

A Review of Long-Term Organic Comparison Trials in the U.S.

Kathleen Delate, Cynthia Cambardella, Craig Chase, and Robert Turnbull

5.1 INTRODUCTION

As early as 1843 in Rothamsted, England, and 1876 in the Morrow Plots in Illinois, U.S.A., agricultural researchers recognized the importance of documenting the impacts of long-term farming systems on crop productivity, soil quality and economic performance. The link between soil quality and farm viability was well understood, as Andrew Sloan Draper, who was President of the University of Illinois when the Morrow Plots were established, stated prophetically that "The wealth of Illinois is in her soil, and her strength lies in its intelligent development" (University of Illinois [UI], 2015). More recently, long-term organic farming system trials across the U.S. have been established to capture similar information. These long-term crop rotation studies also enable more robust economic analyses of potential profit outcomes as compared to experiments of shorter duration (Delbridge, Coulter, King, Sheaffer, & Wyse, 2011).

This paper examines six of the oldest grain-crop-based organic comparison experiments in the U.S. (Table 1), the goal of which is to demonstrate the unique contributions of each site and the usefulness of these sites in communicating agronomic, as well as environmental and economic outcomes from organic agroecosystems, to both producers and policymakers. Of particular interest to producers is the transition period at these sites: the 36 months between the last application of prohibited synthetic inputs and certified organic status. Long-term cropping systems trials can provide baseline data, monitor trends over time, and evaluate new technology in each system, within the context of sustainability indices (Baldock, Hedtcke, Posner, & Hall, 2014). Each site is categorized based on location (weather), soil type, crops, and organic/conventional management practices, to allow comparisons across sites. Additionally, notations on whether the site is certified-organic or organic-compliant (using organic practices without certification) are included. Recently, organic farmers have argued for organic research that is conducted on certified organic sites to ensure a modicum of equivalency as compared to practitioners' experiences. Thus, rotation treatments that would not qualify for organic certification have been discouraged from future comparisons (e.g., one site described below has changed their 2-yr to a 3-yr organic rotation).

TABLE 5.1 Long-term organic comparison trials in the U.S.

Name of experiment	Date initiated	Comparisons	Main crops	Lead entity and location
Farming Systems Trial (FST)	1981	Conv[1] C-S vs. Org 3 and 4-yr rotations	Corn, soybean, wheat	Rodale Institute Kutztown, Pennsylvania
Sustainable Ag Farming Systems (SAFS)	1988	Conv C, W, S, B and T vs. Org C, W, S, B, T, O	Corn, tomato, wheat, bean, safflower, oat/vetch/pea	University of California Davis, California
Variable Input Crop Management Systems (VICMS)	1989	Conv C-S vs. Org 3 (dropped Org 2) and 4-yr rotations	Corn, soybean, oat, alfalfa	University of Minnesota Lamberton, Minnesota
Wisconsin Integrated Cropping Systems Trials (WICST)	1989	Conv C-S vs. Org 3 and 4-yr rotations	Corn, soybean, wheat, oat, alfalfa	University of Wisconsin-Madison Arligton, Wisconsin
Beltsville Farming Systems Project (FSP)	1996	Conv C-S vs. Org 2, 3 and 6-yr rotations	Corn, soybean, wheat	USDA-ARS Beltsville, MD
Long-Term Agroecological Research (LTAR)	1998	Conv C-S vs. Org 3 and 4-yr rotations	Corn, soybean, oat, alfalfa	Iowa State Universty Greenfield, Iowa

[1]Conv = following conventional practices; Org = following certified organic practices.
C= corn; S=soybean; W=weat; O=oat; B=dry bean; S=safflower; T= tomato

Key among organic practices is the necessity of extended crop rotations and organic-compliant soil amendments to optimize production, with each of these practices affecting soil quality, carbon sequestration, nitrogen cycling, and other associated functions. Soil quality is the main driver of optimal organic crop yields. Management of soil organic matter (SOM) to enhance soil quality and supply nutrients is a key determinant of successful organic farming, which involves balancing two ecological processes: mineralization of carbon (C) and nitrogen (N) in SOM for short-term crop uptake, and sequestering C and N in SOM pools for long-term maintenance of soil quality. The latter has important implications for regional and global C and N budgets, including water quality and C storage in soils. The importance of yield comparisons in long-term studies cannot be overlooked, as Seufert, Ramankutty, and Foley (2012) in their meta-analysis of organic and conventional crop yields recognized that optimal yields are central to sustainable food security, in addition to the range of other ecological, social and economic benefits organic farming can deliver. For example, when reviewing the relative yield performance of organic and conventional farming systems worldwide from studies beginning in 1988, Seufert et al. (2012) documented a 5% to 34% lower yield under organic management, depending upon crop and soil type, along with experience related to effective nutrient and pest management practices.

Several commonalities exist among the long-term experiments selected for this review (Table 1). All are systems-level experiments with rotation treatments derived from organic crop rotations practiced in each specific area. With corn (Zea mays L.) and soybean (Glycine max L.) production comprising 56% of the major crops grown in the U.S. (USDA-NASS, 2011), and wheat (Triticum aestivum L.) the third largest crop, one to three of these major crops are present in the trials discussed, as representative of the U.S. agricultural landscape. Because organic systems are complex in nature, in systems-level experiments, the abiotic and biotic components (structure) of the system can be evaluated in terms of the effects on system function (Drinkwater, 2002). Resulting system function data is then used to elucidate factors underlying less than optimal yields (Seufert et al., 2012) and help fine-tune best management practices to improve organic systems.

5.2 THE FARMING SYSTEMS TRIAL (FST) RODALE INSTITUTE, PENNSYLVANIA

The Farming Systems Trial (FST) at Rodale Institute (RI) is the longest-running comparison of organic and conventional agriculture systems in the U.S. Located near Kutztown, Pennsylvania, the soil type is a moderately well-drained Comly silt loam. Established in 1981, in the year following the release of the first comprehensive study of organic agriculture by the USDA, which advocated such comparisons (USDA, 1980; Youngberg & Demuth, 2013), the FST compares two organic systems with a conventional system, using 0.17-ha plots in eight replications, with each crop in the rotation grown every year (Rodale Institute [RI], 2011). The farming systems chosen were based on typical grain crops grown in Pennsylvania: in the conventional system, corn and soybean were grown for 23 years, then wheat was added to the rotation starting in 2004. The two organic systems consisted of corn, soybean, wheat, and red clover (Trifolium pretense L.)-alfalfa (*Medicago sativa L.*) hay in the rotation, and compared contrasting methods for maintaining soil fertility: 1) legume cover crops only, vs. 2) manure-based fertility with cover crops. The conventional system followed land-grant university recommendations for synthetic chemical nutrient and pest management inputs.

The FST was one of the first research units to report on the "transition effect" (Liebhardt et al., 1989), where organic grain yields matched conventional yields after an initial yield decline during the transition years. In 2008, genetically modified (GM) crops and glyphosate-based no-till treatments were added to the conventional comparison, in response to public pressure to compare more current conventional systems. Although organic plots could not be certified organic due to inadequate distance from GM crops, the organic systems always adhered to organic-compliant practices.

While many in the organic community were opposed to RI adding GM crops in the FST, it has been interesting to note that, even with this advanced technology, conventional yields have not improved over non-GM conventional crops, contrary to what proponents believed would occur (RI, 2011). In addition, organic systems have demonstrated greater resiliency during drought, when organic corn yielded 8,411 kg ha-1 compared to 6,403 kg ha-1 in the conventional system (Lotter, Seidel, & Liebhardt, 2003).

The FST was one of the first comparison experiments that monitored water quality, through an underground lysimeter system, and found that leachate from the conventional system more frequently exceeded the NO3-N drinking water standard of 10 ppm than the organic systems (Pimentel, Hepperly, Hanson, Douds, & Sidel, 2005).

The RI also conducted a detailed energy analysis, which included the energy used in the manufacture, transportation and application of fertilizers and pesticides in each FST system. Their analysis identified that FST organic systems consumed 45% less energy than the conventional systems, with N fertilizer composing the largest conventional system energy input at 41% of total energy consumption. Thus, production efficiency was 28% higher in the organic system, with the conventional no-till system having the lowest efficiency, based on high-energy requirements for input manufacturing. In a concomitant analysis, greenhouse gas (GHG) emissions associated with the conventional systems were 40% greater per volume of production than the organic systems (RI, 2011).

Soil health, one of the key attributes in agriculture promoted by RI research, was shown to be greatest in the organic system where manure fertilization was employed, followed by the organic legume system. Annual carbon (C) increases were 981 kg C ha-1 in the organic/manure system, 574 kg C ha-1 in the organic/legume system, and 293 kg C ha-1 in the conventional system (Pimentel et al., 2005). Based on the higher soil quality promoting similar yields to the conventional system, the organic system has proven to be economically competitive, with an analysis conducted by Hanson and Musser (2003) showing only a 10% organic premium price was needed to ensure parity with the conventional system. When prevailing organic price premiums were added, the organic system returns averaged 2.9 to 3.8 times the conventional system (Moyer, 2013). Organic price premiums should be included in economic analyses, as they represent the reward organic farmers reap when practicing organic farming—a premium organic consumers are willing to pay in support of farmers who utilize less environmentally harmful methods of farming (Lin, Smith, & Huang, 2008).

5.3 THE SUSTAINABLE AGRICULTURE FARMING SYSTEMS PROJECT (SAFS), DAVIS, CALIFORNIA

The Sustainable Agriculture Farming Systems project (SAFS) was established in 1988 at the University of California, Davis, to study the transition from conventional to low-input and organic crop production practices (University of California [UC], 2015). The experiment was unique in its study of Mediterranean crops, growing on Reiff loams (coarse-loamy, mixed, non-acid thermic Mollic Xerofluvents) and Yolo silt loams (fine-silty, mixed, non-acid, thermic Typic Xerothents). The SAFS site was located in the state with the highest number of organic farmers in the U.S., which led to the integral role of farmers and farm advisors in the planning, execution, and interpretation of results for greater dissemination to the organic farming community. In addition, organic plots were certified organic by California Certified Organic Farmers (CCOF), a critical factor in the site's applicability for regional farmers. Treatments included two conventional systems: a 2-yr (conv-2) and 4-yr (conv-4) crop rotation; and two 4-yr low-input and organic crop rotations (Poudel et al., 2001). The three 4-yr rotations included tomato, safflower, bean, and corn, while the conv-2 system was a tomato-wheat rotation. In the low-input and organic treatments, an oat/vetch/pea mixture was also part of the rotation. Four replications of each treatment and all crop rotation entry points were planted in 0.12 ha–plots, arranged in a randomized block, split-plot design. Furrow irrigation was used for all systems, typical of farming operations in California. Animal manure and winter cover crops provided fertility in the organic system, while the conventional systems received synthetic fertilizer inputs. The inclusion of a low-input system in long-term organic comparison trials can be problematic (unless it is the sole conventional comparator), because few, if any, of the "low-input" systems follow an equivalent pattern of input applications to allow comparisons across regions. For example, the SAFS low-input system used cover crops and animal manure during the first 3 years, then switched to cover crops and synthetic fertilizer, which would render it as essentially a conventional treatment.

Soils research at SAFS resulted in significant gain in our understanding of the processes involved in enhanced soil quality resulting from organic practices, including increased storage of plant nutrients and C, a reduction in soil-borne diseases, increased pools of P and K, higher microbial biomass and activity, an increase in mobile humic acids and soil water-holding capacity (Clark, Horwath, Shennan, & Scow, 1998). The SAFS site was one of the first experiments to examine soil microbial abundance and activity and determine the importance of cover crops and fall irrigations in promoting bacterial-feeding nematode populations and N mineralization (Jaffee, Ferris, & Scow, 1998), which led to improved organic tomato yields. Additionally, adjustments of grass/legume cover crop mixtures according to soil fertility conditions, along with rotating cover crops, helped prevent stem and foliar diseases. The inclusion of winter cover crops in the low-input and organic systems was a key factor in the success of these systems by supplying soil nutrients and aiding in water infiltration, which proved problematic under conventional management. Suppression of the root-knot nematode, *Meloidogyne javanica*, was associated with high levels of microbial biomass observed in the systems using cover crops (Bossio, Scow, Gunapala, & Graham, 1998). The conventional systems were the least efficient at storing N inputs, which are critical for long-term fertility maintenance (Clark et al., 1998). Microbial community variables were positively correlated with mineral N in the organic system, while the opposite was observed in the conv-4 system (Gunapala & Scow, 1998).

Under California's often challenging climate, organic crops with high N demands, such as tomato and corn, were more susceptible to yield losses compared to conventional and low-input systems receiving annual applications of synthetic N fertilizer, while organic bean and safflower crops produced comparable yields (UC, 2015). As with the FST economic analysis, the importance of premium prices for economic viability was demonstrated, where, among the 4-yr rotations in the SAFS study, the organic system with premium prices was the most profitable (Clark, Klonsky, Livingston, & Temple, 1999). Interestingly, while the low-input system outperformed the organic system agronomically, because of the conventional prices received for low-input crops, this system fell below the two conventional systems in profitability.

In 2002–2003, SAFS began a second phase to examine the interaction of tillage effects on the three historical systems, and explore off-farm environmental quality by joining the Long Term Research on Agricultural Systems (LTRAS) project (UC, 2015). Many in the academic community were disappointed about the loss of such a valuable, long-term certified organic site as SAFS. The history of the SAFS site illustrates the fragility of long-term comparisons absent a strong and enduring institutional commitment. While important information may be derived from the LTRAS site, the LTRAS site does not have the same history of organic farmer involvement and oversight that the SAFS site invited, and many feel is critical for the success of long-term organic sites. As stated on the SAFS website: "Ideas that were once considered to be impractical or even radical are now gaining in popularity. As consumer demand for organic foods increases more growers are considering the transition to organic farming systems and seek out the SAFS project to get information and advice" (UC, 2015).

5.4 THE WISCONSIN INTEGRATED CROPPING SYSTEMS TRIAL (WICST), ARLINGTON, WISCONSIN

The WICST was established in 1989 but, because of a staggered start, every crop phase was not present every year for all the crop rotations until 1992 (Posner, Casler, & Baldock, 1995). Four replications of each crop phase were planted on 0.3-ha plots. The main soil type is a well-drained Plano silt loam (fine-silty, mixed, superactive, mesic Typic Argiudoll). The treatments include six cropping systems (CS): 1) conventional continuous corn (CS1: CC); 2) conventional corn–soybean (CS2: C-S); 3) organic corn–soybean–winter wheat with frost-seeded red clover (CS3: C-S-W/RC); 4) conventional corn–alfalfa (CS4: C-A); 5) organic corn–alfalfa–oat (*Avena sativa L.*) plus field pea (*Pisum sativum L.*) mix, followed by a year of alfalfa hay (CS5: C-A/O/P-A); and a rotationally grazed pasture (CS6: RC/T/BG/OG) seeded to a mixture of red clover, timothy (*Phleum pratense L.*), brome grass (*Bromus inermis L.*) and orchardgrass (*Dactylis glomerata* L.). Soil changes at this site have not been as consistent as other long-term sites, primarily because of a history of a dairy–forage cropping system of corn and alfalfa with manure returned to the land for 20 years before establishing the trial, leading to high organic matter levels (47 kg g-1 at 0–15 cm) prior to the start of the experiment. The most salient observation from the WICST has been the correlation between weather, weeds and organic crop yields (Posner, Baldcock, & Hedtcke, 2008). Because mechanical weed cultivation in organic systems is dependent on dry weather, in the years when wet weather prevented timely weed management, organic corn yields ranged from 72 to 84% of conventional corn yields, and organic soybean yields ranged from 64 to 79% of conventional soybean yields.

Gaining experience and more advanced equipment for organic operations may have also impacted yield differences, as systems nearly equalized when better technology was introduced in the organic systems, and all cropping systems produced positive, average corn yield trends ranging from 0.1 to 0.2 Mg ha-1 yr-1 (Baldock et al. 2014). Similar to the FST results, adding GM crops did not improve yields. This was the first long-term trial to demonstrate that organic forage crop yields were equal or greater than conventional counterparts, with quality sufficient to produce an equivalent volume of milk as the conventional systems (Posner, Baldock & Hedtcke, 2008).

5.5 THE VARIABLE INPUT CROP MANAGEMENT SYSTEMS (VICMS) TRIAL, LAMBERTON, MINNESOTA

The Variable Input Crop Management Systems (VICMS) trial was established in 1989 at the University of Minnesota Southwest Research and Outreach Center near Lamberton, MN. Soil types at the site include Normania clay loam (fine-loamy, mixed, superactive, mesic Aquic Hapludolls), Revere clay loam (fine-loamy, mixed, superactive, mesic Typic Calciaquolls), Ves clay loam (fine-loamy, mixed, superactive, mesic Calcic Hapludolls), and Webster clay loam (fine-loamy, mixed, superactive, mesic Typic Endoquolls) (Porter et al. 2003). Two crop rotations and four management strategies are included in the trial, resulting in eight distinct crop management systems. The original crop rotations were a 2-yr corn-soybean rotation, and a 4-yr corn–soybean-oat/alfalfa–alfalfa rotation. The management strategies are zero-external-input (ZEI), low-external-input (LEI), high-external-input (HEI), and organic-inputs (OI). Liquid swine or beef manure was the external nutrient source in the 2- and 4-yr OI systems (applied at 129-138 kg N ha-1). Treatments were replicated three times in a split-plot arrangement, with main plots as crop rotation, and all phases of each rotation present in each year. Split plots, constituting management systems, are 0.16–ha. As previously mentioned, the original 2-yr organic rotation has been replaced with a 3-yr rotation of corn-soybean-wheat/red clover to align the study more closely with predominant organic crop rotations in the region. From 1992 to 2007, corn grain yield was not reduced in LEI and OI 4-yr rotations compared to the HEI 2-year rotation (Coulter, Delbridge, King, Allan & Schaeffer, 2013). Highest organic corn yields, as observed in other long-term sites, were associated with timely weed management. The benefit of the longer organic rotation was observed with soybean yield response, as the relative soybean yield as a percentage of the HEI 2-yr rotation was greatest in the OI 4-yr rotation from 1992 to 2003 (65%) and in the OI 2- and 4-yr rotations from 2004 to 2007 (38 and 41%, respectively) (Coulter et al., 2013).

Soil quality increased in the organic systems in a similar pattern as other long-term sites. The OI system contained the greatest amount of particulate organic matter and potentially mineralizable C compared to the other systems in both rotations (Coulter et al., 2013). Total soil organic C and microbial biomass was higher in the 4-yr OI system than the 4-yr HEI system. Some of the most important contributions from the VICMS site included a detailed economic analysis of the organic systems, including risk analysis. Delbridge et al. (2011) found that when organic price premiums were applied, the average net return of the organic rotation was considerably larger than that of both conventional rotations ($1329 ha-1 vs. an average of $761 ha-1). Across years and crops, net return was 88% greater with the OI 4-yr rotation than the HEI 2-yr rotation. Organic systems also were found to be stochastically dominant to conventional rotations at all levels of risk aversion (Delbridge, Fernholz, King, & Lazarus, 2013).

5.6 USDA-ARS (AGRICULTURAL RESEARCH SERVICE)-FARMING SYSTEMS PROJECT (FSP)

The FSP was established in 1996 at the USDA-ARS Henry A. Wallace Beltsville Agricultural Research Center (BARC) in Beltsville, Maryland. In contrast to other sites, the FSP was designed to evaluate the sustainability of organic rotations, using typical tillage regimes, compared to conventional cropping systems using both tilled and no-till operations (Cavigelli, Teasdale & Spargo, 2013). Farmers, extension agents, agribusiness professionals, and agricultural researchers were involved in system design. The FSP is comprised of five cropping systems: 1) conventional no-till (NT) corn–soybean–wheat/double-crop soybean rotation: NT: C–S–W/S; 2) a conventional chisel-till (CT) corn–soybean–wheat/soybean rotation: CT: C–S–W/S; 3) a 2-year organic corn–soybean rotation (Org2: C–S); 4) a 3-yr organic corn–soybean–wheat rotation (Org3: C–S–W); and 5) a 6-yr organic corn–soybean–wheat– alfalfa (3 years) rotation (Org6: C–S–W–A–A–A). All plots are 0.1 ha in size and all are managed using full-sized farming equipment. Soils at the site range from poorly-drained to well-drained Ultisols.

Results observed at the FSP support the association between system stability and diversity, with lengthening rotations improving agronomic, economic and environmental performance. Specifically, N availability was greater in the 6-yr organic rotation and yields were greater than the 3-yr organic rotation and 2-yr conventional C-S yields.

Regarding other aspects of soil quality, POMN and SOC in all organic systems were greater than in the conventional NT, which signaled the first report of this phenomenon. Conventional no-till farming, which relies on petroleum-based glyphosate herbicide, is advocated throughout the U.S. for its soil quality enhancement, but the N mineralization potential of the organic system at the FSP was, on average, 34% greater than conventional NT after 14 years. Total potentially mineralizable N in organic systems (average 315 kg N ha-1) was significantly greater than the conventional systems (average 235 kg N ha-1) (Spargo, Cavigelli, Mirsky, Maul & Meisinger, 2011). The SOC was greater in the 6-yr organic rotation compared to NT at all depths except 0 to 2 inches.

Despite the use of tillage in organic systems, soil combustible C and N were higher after 9 years in an organic system that included cover crops compared with the three conventional no-till systems, two of which included cover crops, suggesting that organic practices can potentially provide greater long-term soil benefits than conventional no-till (Teasdale, Coffman & Mangum, 2007). Weed pressure decreased with longer rotations (Teasdale & Cavigelli, 2010), suggesting an allelopathic or competitive effect from multiple years of alfalfa–a solid-seeded crop that was cut regularly, which inhibited weed growth. Economic risk also decreased as rotation length increased, and organic returns averaged $706 ha-1 compared to $193 ha-1 (Cavigelli, Hima, Hanson, Teasdale, Conklin, & Lu, 2009). Throughout the mid-Atlantic states, rising concerns regarding nitrate and phosphate fertilization pollution into fragile waterways, like the Chesapeake Bay, has led to increasing restrictions and research on pollution-mitigation methods. A beneficial outcome of the 6-yr organic rotation in this regard was that less poultry manure was needed for optimal yields compared to shorter rotations, thus decreasing nitrate and phosphate pollution potential.

5.7 THE LONG-TERM AGROECOLOGICAL RESEARCH (LTAR) EXPERIMENT, IOWA

The LTAR experiment was established in 1998 at the Iowa State University Neely-Kinyon Farm in Greenfield, Iowa, with funding from the Leopold Center for Sustainable Agriculture. This support allowed focus groups of conventional and organic farmers to help determine the appropriate design and purpose of the LTAR experiment (Delate & DeWitt, 2004). Farmers requested a long-term comparison of the ecological and economic outcomes of conventional and organic cropping systems. The research was then constructed to evaluate alternatives to the traditional corn–soybean rotation in Iowa, and investigate production processes based on agroecological principles, designed to reduce off-farm energy demand and to increase the internal resilience of agroecosystems, which consequently increases their adaptability to potential climate change. Unlike purely research-based experiments, the goal of the LTAR site is to encourage transition to organic production, by documenting the environmental services in organic systems that contribute to climate change mitigation and enhancement of soil quality, crop health, productivity, and food quality. Objectives include identifying cropping systems within the LTAR experiment that maximize yields and soil quality, by fostering carbon sequestration and minimizing nutrient loss; promoting supporting and provisioning ecosystem services of biodiversity, pest suppression, water quality, and soil health through the integration of C-stabilizing components; increasing economic returns by reducing costs of production in field operations and labor, decreasing dependence on external sources of applied fertility, lowering energy costs, and gaining carbon credits.

Finally, educational objectives include field days, workshops and pasture walks for farmers, students, and agricultural professionals to increase understanding and facilitation of the transition to organic production.

The LTAR experiment is located on a 7-ha ridge top with a uniform slope of 0 to 2% with the predominant soil type a moderately well-drained Macksburg silty clay loam (fine, smectitic, mesic Aquic Argiudolls). The cropping system treatments at the LTAR site were designed based on local farmer input with the goal of organic certification 36 months after establishment. Each crop in each rotation is replicated four times in 0.1–ha plots.

Rotations include: 1) conventional corn-soybean (C–S); 2) organic corn-soybean-oats/alfalfa (C–S–O/A); and 3) organic corn-soybean-oats/alfalfa-alfalfa (C–S–O/A–A). Conventional crops are maintained with synthetic fertilization and pesticides, while certified organic fertilization and pest management methods are used in organic plots, using typical farming equipment for the area. Effects of system and rotation treatments are determined for crop productivity and yields; weed, insect, disease, and nematode pest management; soil quality and fertility; nutrient retention and balance; and grain quality. Economic analyses, determined for each treatment, include costs of inputs, subsequent yields, and selling price of organic and conventional crops. Over 13 years, LTAR organic corn and soybean yields were equivalent or greater than conventional counterparts.

Unlike many studies where organic yields suffer during the transition phase, the first LTAR transitioning-to-organic phase demonstrated corn yields in the organic system that were 92% of conventional corn yields while organic soybean yields were 99.6% of conventional soybean yield (Delate & Cambardella, 2004). The advantage of the longer, 4-year organic rotation, which included two years of a perennial legume crop, was exhibited by organic corn yields that averaged 99% of the average conventional corn yield in the post-transitioning phase (Delate, Cambardella, Chase, Johanns, & Turnbull, 2013). Organic soybean yields were 5% greater in the organic rotations than conventional soybean yields. Soil quality results from the LTAR showed that overall soil quality, and especially soil N mineralization potential, was highest in the 4-year organic crop rotation. The organic soils had more soil organic carbon, total N, microbial biomass C, labile organic N, higher P, K, Mg and Ca concentrations, and lower soil acidity than conventional soils. A particularly interesting soils result was obtained in 2012, when an extended drought period was experienced, with 22 cm below normal rainfall during the growing season, and an average of 3 °C above normal temperatures in July. At the end of the 2012-growing season, particulate organic matter

C (POM-C) was higher in the organic soils than the conventional, likely because of altered rates of decomposition of new residue C inputs during this especially dry year (Table 5.2). Soil quality enhancement was particularly evident for labile soil C and N pools, which are critical for maintenance of N fertility in organic systems, and for basic cation concentrations, which control nutrient availability through the relationship with cation exchange capacity (CEC). Despite the serious drought conditions during the growing season in 2012, organic management enhanced agroecosystem resilience and maintained a critical soil function, the capacity to supply nutrients to the crops. Carbon budgets developed after 10 years of organic production showed that the 4-yr organic cropping system can potentially sequester as much soil organic carbon (SOC) in the top 15 cm as obtained when converting from plowing to no-tillage, which is considered the best management practice in conventional farming.

Economic returns mirrored those previously reported at other sites, with the organic rotations garnering, on average, twice the returns of the conventional rotation (Delate et al., 2013), and lower costs than conventional crops during transition (Delate, Duffy, Chase, Holste, Friedrich, & Wantate, 2003; Delate, Chase, Duffy, & Turnbull, 2006). Results from the LTAR experiment have been similar to other long-term trials, although LTAR organic yields have often exceeded those reported in the literature. Higher than usual yields during the transition phase could be attributed to the overall fertility of the Mollisols at the site and the high level of weed management experience, which has been a key aspect of success. Despite the equivalence in net C input, the soil under organic management holds significantly more C than the soil under conventional management, and over the coming decade, we will continue to monitor resulting changes in soil edaphic and biotic characteristics including soil microbial community structure and function under the various cropping systems.

5.8 CONCLUSIONS

The six long-term organic comparison sites examined in this review have contributed to an invaluable understanding of the mechanisms underpinning higher soil quality in organic systems, particularly enhanced C and N storage, leading to competitive economic returns. All experiments were transdisciplinary in nature; analyzed comprehensive system components (productivity, soil health, pest status, and economics); and contained all crops within each rotation and cropping system each year, a critical factor for analysis across years.

TABLE 5.2 Neely-Kinyon Long-Term Agroecological Research (LTAR) experiment soil quality–Fall 2012

	SOC	TN	POM-C	POM-N	MBC	PMIN-N	NO_3-N	P	K	Mg	Ca	pH	Aggs%	BD
	gkg^{-1}	gkg^{-1}	gkg^{-1}	gkg^{-1}	$mgkg^{-1}$	$mgkg^{-1}$	$mgkg^{-1}$	$mgkg^{-1}$	$mgkg^{-1}$	$mgkg^{-1}$	$mgkg^{-1}$			gcm^{-3}
ConvC-S[1]	23.1	2.4	3.0	0.31	275	40.1	21.4	21.2	185	366	3487	6.09	34.9	1.27
OrgC-SO/A	25.7	2.6	4.5	0.33	270	51.9	20.5	57.5	283	413	3870	6.51	35.0	1.22
OrgC-S-O/A-A	24.8	2.5	3.8	0.23	296	52.1	19.7	34.0	251	407	3831	6.41	41.2	1.21
OrgC-S-C-O/A	24.7	2.5	4.3	0.28	362	52.2	16.7	27.4	203	479	3866	6.34	45.4	1.13
$LSD_{0.05}$	1.4	0.1	1.1	NS	42	7.1	NS	12.7	50.9	50.1	161	0.19	7.4	0.08

[1]Results from five randomly-located soil cores (0-15 cm), composited, and removed from each plot after fall harvest, prior to any tillge. Conv = conventionl; Org = certified organic; C = corn; S = soybean; O = oats; and A = alfalfa. SOC = soil organic carbon; TN = total nitrogen; POM-C = particulate organic matter-carbon; POM-N = particulate organic matter-nitrogen; MBC = microbial biomass carbon; NO3-N = nitrate-nitrogen; P = phosphorus; K = potassium; Mg = magnesium; Ca = calcium; Aggs = aggregate stability; BD = bulk density; LSD = Least Significant Difference at $p<0.05$; NS = not significant.

Plot size ranged from 0.1 to 0.3 ha–an area of sufficient size to utilize farm-scale machinery and provide an accurate portrayal of typical farmer experience–often lacking in research station plot research. While, ideally, on-farm sites with larger fields should be employed as comparators to field station experiments to allow a minimum comparison of 5 to 10 years since conversion from conventional farming, as promulgated by Sir Albert Howard (1946), oftentimes, long-term on-farm sites are difficult to obtain and manage. Comparisons with organic grain yields reported from organic farmer surveys in Iowa showed a reduction of 17-20% in organic corn and soybean yields, but returns comparable to the 2X results demonstrated in the long-term trials (Chase, Delate, Liebman, & Leibold, 2008). Organic yield performance was improved in four of the six sites with increased experience and timely weed management, while two sites (FSP and LTAR) with experienced farm managers reported adequate weed control and concomitant equivalent organic and conventional yields early in the long-term site's history. The addition of manure, along with legume forages/cover crops, in the organic fertility scheme has proven essential for sufficient soil quality to support optimal yields across all sites. The scientific rigor under which these sites were operated has provided strong evidence supporting the viability of organic cropping systems for farmers and policymakers alike. Wherever organic farmer involvement in experimental design and feedback was explicit, and organic certification was obtained, organic comparison sites appeared to be more successful in terms of engagement and dissemination of results.

With organic product supply lagging behind the expanding market demand, partially owing to the perceived obstacles to successful transition to organic production (Dimitri & Oberholtzer, 2009), these sites provided sufficient evidence of the potential for successful organic transition. Adoption of land management strategies that foster C sequestration in agricultural soils will be important over the next several decades as we develop new mitigation strategies and technologies to reduce C emissions (Smith, 2004). Agricultural land management options currently recommended to foster C sequestration nearly always include some reduction in tillage intensity, which has been the on-going, or second-phase research, of four of the long-term sites (FST, SAFS, FSP; now VICMS), and implementation of integrated, multifunctional cropping rotations that include cover crops/forage legumes, small grains, and animal manure/compost soil amendments, as demonstrated by all long-term sites. Water quality enhancement, by reducing NO3-N loss through the adoption of organically managed extended rotations that include small grains, forage legumes and pasture (Cambardella & Delate), is considered an integral part of the next phase of many of the long-term trials. These results suggest that organic farming practices have the potential to reduce nitrate leaching, foster carbon sequestration, and allow farmers to remain competitive in the marketplace. Institutional support for these long-term comparisons is critical for successful organic farming demonstrations for area farmers and policymakers.

REFERENCES

1. Baldock, J. O., Hedtcke, J. L., Posner, J. L., & Hall, J. A. (2014). Organic and conventional production systems in the Wisconsin Integrated Cropping Systems Trial: III. Yield trends. Agronomy Journal, 106, 1509-1522. http://dx.doi.org/10.2134/agronj14.0004
2. Bossio, D. A., Scow, K. M., Gunapala, N., & Graham, K. J. (1998). Determinants of soil microbial communities: Effects of agricultural management, season, and soil type on phospholipid fatty acid profiles. Microbial Ecology, 36, 1-12. http://dx.doi.org/10.1007/s002489900087
3. Cavigelli, M., Hima, B., Hanson, J., Teasdale, J., Conklin, A., & Lu, Y. (2009). Long-term economic performance of organic and conventional field crops in the mid-Atlantic region. Renewable Agriculture and Food Systems 24, 102-119. http://dx.doi.org/10.1017/S1742170509002555
4. Cavigelli, M. A., Teasdale, J. R., & Spargo, J. T. (2013). Increasing crop rotation diversity improves agronomic, economic, and environmental performance of organic grain cropping systems at the USDA-ARS Beltsville
5. Farming Systems Project. Crop Management.
6. Chase, C., Delate, K., Liebman, M., & Leibold, K. (2008). Economic analysis of three Iowa rotations. PMR 1001. Iowa State University, Ames, IA.
7. Clark, M. S., Horwath, W. R., Shennan, C., & Scow, K. M. (1998). Changes in soil chemical properties resulting from organic and low-input farming practices. Agronomy Journal, 90, 662-671. http://dx.doi.org/10.2134/agronj1998.00021962009000050016x
8. Clark, S., Klonsky, K., Livingston, P., & Temple, S. (1999). Crop-yield and economic comparisons of organic, low-input, and conventional farming systems in California's Sacramento Valley. American Journal of Alternative Agriculture, 14, 109-121. http://dx.doi.org/10.1017/S0889189300008225
9. Coulter, J. A., Delbridge, T. A., King, R. P., & Sheaffer, C. C. (2013). Productivity, economic, and soil quality in the Minnesota Variable-Input Cropping Systems Trial. Crop Management.
10. Delate, K., & Cambardella, C. (2004). Agroecosystem performance during transition to certified organic grain production. Agronomy Journal, 96, 1288-1298. http://dx.doi.org/10.2134/agronj2004.1288
11. Delate, K., Cambardella, C., Chase, C., Johanns, A., & Turnbull, R. (2013). The Long-Term Agroecological Research (LTAR) experiment supports organic yields, soil quality, and economic performance in Iowa. Crop Management. http://dx.doi.org/10.1094/CM-2006-1016-01-RS

12. Delate, K., Chase, C., Duffy, M., & Turnbull, R. (2006). Transitioning into organic grain production: An economic perspective. Crop Management.
13. Delate, K., & DeWitt, J. (2004). Building a farmer-centered land grant university organic agriculture program—A Midwestern partnership. Renewable Agriculture and Food Systems, 19, 1-12. http://dx.doi.org/10.1079/RAFS200065
14. Delate, K., Duffy, M., Chase, C., Holste, A., Friedrich, H., & Wantate, N. (2003). An economic comparison of organic and conventional grain crops in a Long-Term Agroecological Research (LTAR) site in Iowa. American Journal of Alternative Agriculture, 18, 59-69. http://dx.doi.org/10.1079/AJAA200235
15. Delbridge, T. A., Coulter, J. A., King, R. P, Sheaffer, C. C., & Wyse, L. (2011). Economic performance of long-term organic and conventional cropping systems in Minnesota. Agronomy Journal, 103, 1372-1382. http://dx.doi.org/10.2134/agronj2011.0371
16. Delbridge, T. A., Fernholz, C., King, R. P., & Lazarus, W. (2013). A whole-farm profitability analysis of organic and conventional cropping systems. Agricultural Systems. http://dx.doi.org/10.1016/j.agsy.2013.07.007
17. Dimitri, C., & Oberholtzer, L. (2009). Marketing US organic foods: Recent trends from farms to consumers. USDA-ERS, Washington, DC.
18. Drinkwater, L. E. (2002). Cropping systems research: Reconsidering agricultural experimental approaches. HortTechnology, 12, 355-361.
19. Gunapala, N., & Scow, K. M. (1998). Dynamics of soil microbial biomass and activity in conventional and organic farming systems. Soil Biology and Biochemistry, 30, 805-816. http://dx.doi.org/10.1016/S0038-0717(97)00162-4
20. Hanson, J. C., & Musser, W. N. (2003). An economic evaluation of an organic grain rotation with regards to profit and risk. Working Paper 03-10. Department of Agricultural and Resource Economics, University of Maryland, College Park, MD. http://dx.doi.org/10.1596/1813-9450-3168
21. Howard, A. (1946). The War in the Soil. Rodale Press. Emmaus, PA.
22. Jaffee, B. A., Ferris, H., & Scow, K. M. (1998). Nematode-trapping fungi in organic and conventional cropping systems. Phytopathology, 88, 344-350. http://dx.doi.org/10.1094/PHYTO.1998.88.4.344
23. Liebhardt, W., Andrews, R., Culik, M., Harwood, R., Janke, R., Radke, J., &Rieger-Schwartz, S. (1989). Crop production during conversion from conventional to low-input methods. Agronomy Journal, 81, 150-159. http://dx.doi.org/10.2134/agronj1989.00021962008100020003x
24. Lin, B-H., Smith, T. A., & Huang, C. L. (2008). Organic premiums of U.S. fresh produce. Renewable Agriculture and Food Systems, 23(3), 208-216. http://dx.doi.org/10.1017/S1742170508002238
25. Lotter, D., Seidel, R., & Liebhardt, W. (2003). The performance of organic and conventional cropping systems in an extreme climate year. American Journal of Alternative Agriculture, 18, 146-154. http://dx.doi.org/10.1079/AJAA200345
26. Moyer, J. (2013). Perspective on Rodale Institute's Farming Systems Trial. Crop Management. http://dx.doi:10.1094/CM-2013-0429-03-PS
27. Pimentel, D., Hepperly, P., Hanson, J., Douds, D., & Sidel, R. (2005). Environmental, energetic, and economic comparisons of organic and conventional farming systems. BioScience 55, 573-582. http://dx.doi.org/10.1641/0006-3568(2005)055%5B0573:EEAECO%5D2.0.CO;2

28. Porter, P. M., Huggins, D. R., Perillo, C. A., Quiring, S. R., & Crookston, R. K. (2003). Organic and other strategies with two and four year crop rotations in Minnesota. Agronomy Journal, 95, 233-244. http://dx.doi.org/10.2134/agronj2003.0233
29. Posner, J. L., Baldock, J. O., & Hedtcke, J. L. (2008). Organic and Conventional Production Systems in the Wisconsin Integrated Cropping Systems Trials: 1. Productivity 1990-2002. Agronomy Journal, 100, 253-260. http://dx.doi.org/10.2134/agrojnl2007.0058
30. Posner, J. L., Casler, M. D., & Baldock, J. O. (1995). The Wisconsin integrated cropping system trial: Combining agro-ecology with production agronomy. American Journal of Alternative Agriculture, 10, 98-107. http://dx.doi.org/10.1017/S0889189300006238
31. Poudel, D., Ferris, H., Klonsky, K., Horwath, W. R., Scow, K. M., van Bruggen, A. H. C., ... Temple, S. R. (2001). The sustainable agriculture farming system project in California's Sacramento Valley. Outlook on Agriculture, 30, 109-116. http://dx.doi.org/10.5367/000000001101293553
32. Rodale Institute. (2011). The Farming Systems Trial-Celebrating 30 years. Rodale Institute, Kutztown, Pennsylvania.
33. Seufert, V., Ramankutty, N., & Foley, J. A. (2012). Comparing the yields of organic and conventional agriculture. Nature, 485, 229-232. http://dx.doi.org/10.1038/nature11069
34. Smith, P. (2004). Carbon sequestration in croplands: The potential in Europe and the global context. European Journal of Agronomy, 20, 229-236. http://dx.doi.org/10.1016/j.eja.2003.08.002
35. Spargo, T. J., Cavigelli, M. A., Mirsky, S. B., Maul, J. E., & Meisinger, J. J. (2011). Mineralizable soil nitrogen and labile soil organic matter in diverse long-term cropping systems. Nutrient Cycling in Agroecosystems, 90, 253-266. http://dx.doi.10.1007/s10705-011-9426-4
36. Teasdale, J. R., & Cavigelli, M. A. (2010). Subplots facilitate assessment of corn yield losses from weed competition in a long-term systems experiment. Agronomy for Sustainable Development, 30, 445-453. http://dx.doi.org/10.1051/agro/2009048
37. Teasdale, J. R., Coffman, C. B., & Mangum, R. W. (2007) Potential long-term benefits of no-tillage and organic cropping systems for grain production and soil improvement. Agronomy Journal, 99, 1297-1305.
38. University of California (UC)–Davis. (2015). Sustainable Agriculture Farming Systems (SAFS) Website. Retrieved from http://safs.ucdavis.edu
39. University of Illinois (UI). (2015). The Morrow Plots: A Century of Learning, Department of Crop Sciences, UI, Urbana, IL. Retrieved from http://cropsci.illinois.edu/research/morrow
40. U.S. Department of Agriculture (USDA)–National Agricultural Statistics Service (NASS). (2011). Crop Production–2011. Retrieved from http://www.nass.usda.gov
41. U.S. Department of Agriculture (USDA) Study Team on Organic Agriculture. (1980). Report and Recommendations on Organic Farming. U.S. Department of Agriculture, Washington, D.C. Retrieved from http://www.nal.usda.gov/afsic/pubs/USDAOrgFarmRpt.pdf
42. Youngberg, G., & Demuth, S. P. (2013). Organic agriculture in the United States: A 30-year retrospective. Renewable Agriculture and Food Systems, 28, 294-328. http://dx.doi.org/10.1017/S1742170513000173

PART II
Preparing for the Future

CHAPTER 6

Keeping the Actors in the Organic System Learning: The Role of Organic Farmers' Experiments

Christian R. Vogl, Susanne Kummer, Friedrich Leitgeb, Christoph Schunko, and Magdalena Aigner

6.1 INTRODUCTION

6.1.1 THE IMPORTANCE OF INNOVATION

The performance of agriculture worldwide clearly shows that the current mainstream agricultural pathway is not sustainable, causing a diversity of ecological, social and economic problems (McIntyre et al., 2009). Currently, innovation (e.g. Smits et al., 2010; EU SCAR, 2012) is seen as the buzzword and the key concept for supporting the urgently needed pathways or transition towards sustainability (van de Kerkhof & Wieczorek, 2005) or towards resilient societies (Folke et al., 2010; Leach et al., 2012), particularly in organic farming. Organic farming has contributed to multiple aspects of sustainability, especially concerning (new/innovative) on-farm production methods (Darnhofer, 2014a, Moeskops et al., 2014).

For a long period of time the term 'innovating' was mainly associated with science or commercial enterprises. Recently the focus has shifted and clear evidence has been presented, that innovation is a dynamic social multi-stakeholder process that implies the participation of a diversity of stakeholders (Smits et al., 2010). Today, participatory action research (McIntyre, 2007; Chevalier & Buckles, 2013), citizen science (Tulloch et al., 2013) or transdisciplinary research (Tress et al., 2005; Mittelstraß, 2011) are state of the art approaches for ensuring that not only local knowledge, but also creativity and enthusiasm from all stakeholders linked to a certain topic are involved and taken seriously in the related research and innovation pathways.

In the agricultural sciences sector the debate on the role of stakeholders in providing information, sharing knowledge and supporting innovation is far advanced and has been framed in various models like e.g. in the Agricultural Knowledge and Information System or the Agricultural Innovation System, i.e. AKIS or AIS (e.g. Rivera et al., 2005; Spielmann, 2008). Nevertheless, the creative process that leads to farmers' innovations is rarely studied nor described precisely in agricultural sciences e.g. in syntheses on Agricultural Knowledge and Innovation Systems (EU SCAR, 2012) and in policy papers on innovation in organic farming (Moeskops et al., 2014).

The concepts currently used for describing what leads to farmers innovations are e.g. 'problem solving', 'innovating' or 'self help' (Moeskops et al., 2014). These terms are however used ambiguously and imprecisely, which might easily lead to ignoring the complexity of the processes involved. A lack of knowledge of this genuine creative process of 'innovating' might also lead to ignoring the intervening factors, misplacing the key incentives and thus not sufficiently taking into account the opportunities for encouraging farmers' innovations especially in organic farming.

6.1.2 FARMERS' EXPERIMENTS

In this paper we pick up and propose the concept of farmers' experiments as one option for describing the creative process that might lead to farmers' innovations. Yet, an experiment in general is defined as 'a course of action tentatively adopted without being sure of the outcome' (ODO, 2010) or 'a test or series of tests in which purposeful changes are made to the input variables of a process or system so that we may observe and identify the reasons for changes that may be observed in the output response' (Montgomery, 2009). Farmers' experimentation is the process by which farmers informally conduct trials or tests that can result in new knowledge and innovative management systems suitable for their specific agro-ecological, socio-cultural and economic conditions (Rajasekaran, 1999). Sumberg and Okali (1997), who did pioneer work on farmers' experiments, consider two conditions necessary for an activity to be labelled an experiment: the creation and initial observation of conditions, and the observation or monitoring of subsequent results.

6.1.3 LINKS TO ORGANIC FARMING

There are two reasons why it is particularly interesting to explore farmers' experiments in the context of organic farming. First, sustainable land use practices are more knowledge-intensive (Röling & Brouwers, 1999). While conventional farmers can use external inputs such as synthetic pesticides and synthetic fertilisers to handle adverse dynamics in their agro-ecosystem, organic farmers need to develop knowledge about the agro-ecosystem to a larger extent to be able to manage their farms successfully without these inputs.

Second, organic agriculture was developed by farmers and farmers' grassroots organisations. Academic science and research only played a minor role in the historical development of organic agriculture (Padel, 2001), and organic farming was developed by practical experiments and trials of farmers and practical researchers. The lack of advice and formal research in the pioneer phase of organic agriculture leads to the assumption that organic farmers have nurtured a culture of experimentation. Organic farmers in the pioneer phase can be referred to as active experimenters and practical researchers (Gerber et al., 1996).

To our assumption it was not only the pioneers of organic agriculture who experimented; many organic and non-organic farmers worldwide are presumably still actively trying and experimenting to answer questions and solve problems that emerge continuously. We were interested in addressing this assumption in field sites that are very different from each other and assessing if and to which extent organic farmers realize activities that can be called farmers' experiments. We focus on experiments carried out by farmers on their own initiative, and we explicitly avoid referring to on-farm research.

We want to contribute to the current debate on the elements needed for encouraging innovation in organic farming. We do so by presenting empirical evidence from Austria and Cuba that farmers' experiments are a key element of innovating at farm level and by discussing the potential role of farmers' experiments in the innovation process.

6.2 CASE STUDIES AND METHOD

Austria and Cuba were selected for field research (together with Israel; Data on Isreael not presented here) due to various criteria that cause variation between the study sites. Austria has a long history of third party certified organic farming under a formal regulatory and policy framework and is an industrialized country in a temperate climate with high availability of farm inputs and formal advisory on organic farming; Cuba counts with a well organized but relatively young agroecology movement, which is the national interpretation of organic farming, and is a tropical country with limited availability of farm inputs (Kummer et al., 2012) in prep; Leitgeb et al., 2011, 2014).

Field research in Austria and Cuba started with semi-structured interviews (Austria: n = 47; Cuba n = 72; both in 2007 and 2008) based on samples of farmers with maximum variation (criteria for variation e.g.: region, different production types) for learning the terminology and aspects related to the topic of 'changes at farm level' and 'trying something'). Semi-structured interviews were digitally recorded, transcribed with the software ExpressScribe, and processed with the software Atlas.ti. We used qualitative content analysis, employing a combination of deductive and inductive coding for learning on such aspects as the topics, methods, outcomes, attitudes and beliefs related to the process of trying, testing, changing 'something' at farm level. We expected that the term 'experiment' might be loaded with the connotation of a scientific procedure (Sumberg & Okali, 1997, p. 58). It was therefore agreed not to use that term during the semi-structured interviews to prevent narrowing the research field with this specific, technical connotation. The terms we used to refer to experimentation activities during interviews were 'to try, to try something, to try something new' (the terms we used in German were 'etwas probieren' or 'etwas ausprobieren', in Spanish: 'probar algo').

Based on insights from these semi-structured interviews a structured questionnaire was set up with pre-defined answer categories on all elements of experimentation identified. In the structured interviews, in contrast to the semi-structured interviews before, the conversation was started with the purposeful introduction of the term 'experiment', including a definition that was based on the results of the semi-structured interviews. The structured interviews were applied in Austria with 76 organic farmers in 2008 and 2012 and in Cuba with 34 farmers from the Cuban Agroecology Movement and the Cuban Urban agriculture movement in 2007 and 2008.

Structured interviews from Austria were digitally recorded, data inserted into a Microsoft-Access database, and later descriptively analyzed with Microsoft-Excel and SPSS, from Cuba this data set has not yet been analyzed. Here selected qualitative descriptive data is presented summarizing the results from Kummer et al. (2012, 2015 in prep) and Leitgeb et al. (2011, 2014).

6.3 RESULTS

In Austria the interviewees in semi-structured and structured interviews (together n=123) mentioned individual topics for experimenting and only eight interviewed farmers stated that they had never carried out any activity that they would define as 'trying something'. In Cuba 370 individual topics were mentioned by all farmers in semi-structured interviews.

Aspects in crop production (e.g. introduction of new species or varieties) were the most frequently mentioned topics in Austria and Cuba, but literally all aspects are to be found, and even commercialization, construction, testing of alternative remedies or the influence of the lunar cycle, or social organisation were under the topics mentioned by the Austrian and Cuban farmers for doing experiments.

The most frequently mentioned motives for doing experiments were in both countries personal reasons and overcoming challenges or problems. Challenges frequently cited in Cuba were e.g. increasing productivity or achieving independence from external resources. Personal reasons included a general interest in a specific topic or curiosity about how something could work or not, and also the opinion that implementing a specific practice on the farm would be meaningful and desirable for the respective person. Farmers in both countries mentioned most frequently other farmers as sources for information needed for the experiment and also as sources for ideas, together with literature or advisors.

In Austria, two-thirds of the farmers who experimented had an explicit mental or written plan before starting. In Cuba one third of the respondents had precise plans, partly in a written way based upon detailed criteria. In both countries, the majority of the farmers stated that they set up their experiments first on a small scale and enlarged them if the outcome of the experiments was satisfactory. Repetitions were done by running experiments in subsequent years between two and five years long, partly longer but without documenting the duration by our respondents.

In both countries the majority of the farmers monitored the experiments regularly, mostly through observation and comparisons (e.g. with previous experiences, with other farmers, with another unit at the own farm, etc.). Only a small proportion of farmers did measurements. Documentation strategies included taking notes, pictures, samples or videos. In both countries, many experiments were not discrete actions but nested in time and space: One specific experiment can be the source of information or motivation for another specific experiment, experiments can be carried out simultaneously and a 'smaller' topic under experimentation can be part of a 'bigger' topic under experimentation.

In Austria farmers most frequently reported as outcomes of their trials having obtained more knowledge, having learned and increased satisfaction, but also having reduced the workload, increased production, gained reputation or increased income. Increased productivity, increased self sufficiency and better work efficiency were the most frequently mentioned outcomes in Cuba. First addresses for disseminating outcomes were other farmers in both countries. Having learned was attributed by the farmers even to failures or flops in experimenting.

In Cuba, experimentar (experimenting) and experimento (experiment) were terms frequently used by farmers when answering to our questions about 'trying something'. This was different from Austria where *etwas ausprobieren* (to try something) was the most common phrase used. In Cuba, *experimentación campesina* (farmers' experiments) was an integral part of the Cuban agroecology movement and therefore understood by most of the respondents as a concept and as a practical daily activity. Experimentation, innovation and inventions at farm level are part of the Cuban discourse on rural development and encouraged explicitely e.g. through competitions for the best innovation or invention at municipal, provincial and national level, including awarding them for innovations or supporting the negotiations for achieving a patent for promising inventions. In Austria, a formal discourse on farmers' experiments in the organic farming movement or under organic farmers, even when talking about 'to try something' could not be observed during the study period.

6.4 DISCUSSION

With data from Austria and Cuba we can empirically confirm findings of e.g. Sumberg and Okali (1997) that farmers engage in activities of 'trying something'.

These activities can be called farmers' experiments as they include to a considerable proportion planning, implementing variables of unknown consequences in search for their effects, monitoring the effects, and communicating results.

Various authors draw diverse conclusions about the significance of farmers' experiments, but most of the authors agree that all farmers have experimental capacity (e.g. Rhoades & Bebbington, 1991; Chambers, 1999; Quiroz, 1999; Critchley & Mutunga, 2003; Bentley, 2006, 2010), and that experiments are an integral part of farming activities (Sumberg et al., 2003). The experimental capability of farmers, similarly to the resilience of farms, cannot be regarded simply as an automatic response being deducted from the farms' characteristics, but it is rather the ability to identify opportunities, implement options and to 'learn as part of an iterative, reflexive process' (Darnhofer, 2014b).

However, this does not mean that all farmers are innovative (Quiroz, 1999). Experimenting farmers are rarely a homogeneous group. They have been found to be both resource-rich and resource-poor (Saad, 2002), both men and women, both outsiders and well-integrated, and both highly educated and less educated (Reij & Waters-Bayer, 2001). Farmers conduct experiments to test their ideas in their own way (Rajasekaran, 1999).

Experimentation can be induced by intuition, curiosity or by an explicit desire to learn (Stolzenbach, 1999). Farmers can be driven by economic motives as well as by a concern for production, and saving labour or capital (Critchley, 2000; Bentley, 2006). While new ideas and changes spark creativity and induce experiments, the capacity to experiment and learn also depends on prior knowledge and experiences. The source of farmers' experiments is therefore a combination of prior local/traditional knowledge of the farmer and new information the farmer aquires from elsewhere (Bentley, 2006).

Based on our findings and literature, we propose a theoretical model of the experimentation process (Figure 5.1) that helps elucidating what usually just has been vaguely called 'problem solving', 'innovating' or 'self help (Moeskops, 2014). When a certain problem or topic arises, a farmer can decide to adopt an available solution to deal with the situation (Wortmann, 2005), without entering an experimentation process. If the farmer decides to start an experiment, he or she can adapt a common solution that is already known to him or her (Wortmann, 2005; Pretty, 1995), or can decide to try something new. The experimentation process can be defined as a research process that involves a specific methodological approach, including setup, monitoring of the process and evaluation of the results. Different factors, such as environmental, economic or social conditions influence the experimentation process (Sumberg & Okali,

1997), and have an effect on the experiment. Interrelations also exist with regard to the communication system in which the farmer is involved: farmers use local knowledge from their own farm in combination with knowledge from other sources, such as other farmers, media, science or advisory services (Stolzenbach, 1999; Bentley, 2006; Sturdy et al., 2008; Leitgeb et al., 2011). The results of an experimentation process can be classified into innovations, inventions or 'failures' (the later being learning experiences but not involving any change at farm level). These results are usually communicated to the social network of the farmers, such as e.g. family, neighbors or advisors. They are also fed back into the planning and implementation of new experiments to be realized by the farmer.

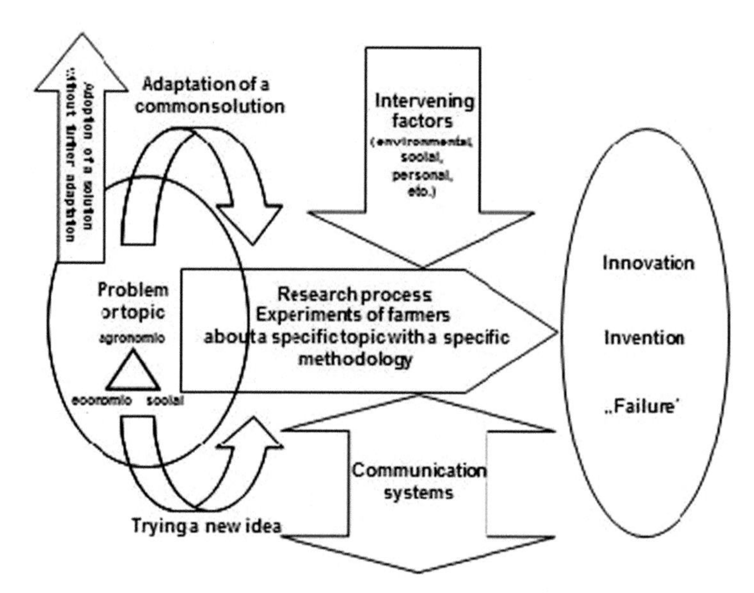

FIGURE 6.1 Model adapted from Ninio and Vogl (2006) for operationalizing the topic of farmers' experiments

Farmers, at least in Austria, themselves hardly use the term 'experiment' to refer to their practical on-farm experiments, but relate this term more to a scientific and formal procedure. In various empirical studies on the topic, using the term 'trying' instead of 'experimenting' in interviews has been seen as being more expedient (Sumberg & Okali, 1997), while in other cases local terms are used to address the subject in the field (Stolzenbach, 1999). Other terms used in literature are i) 'farmer research', which refers to 'research conducted by farmers for discovery or production of information' (Wortmann et al., 2005) and ii) 'on-farm research', which means research conducted, and usually also controlled, by scientists on farms, involving the farmer more or less actively (Lawrence et al., 2007).

The topic of this paper was discussed during various oral presentations and poster presentations with peer scientists. In such discussions we often observed that scientists can be quite reluctant in accepting that the term experiment may be used also by actors not affiliated to academic science, and that e.g. farmers carry out their own experiments. For academic scientists, obvious limitations of farmers' experiments are e.g. precision, reliability, robustness, accuracy, validity or the correct analysis of cause and effect. To our assessment based on our results, but also confirmed e.g. by Moller (Henrik Moller, personal communication, November 2nd, 2014), comparative weakness of the farmers' experiments compared to the formal science experiments often include:

- Lack of or poor spatial and temporal replication;
- Few treatments, usually one at a time;
- Reliance on a 'Before-After' comparison for detection of an experimental effect. Many farmer's experiments reject or accept an innovation/change if it works/doesn't work after a year or two—whereas ecological systems often display time-treatment interaction effects (what works now may not work in a different year and vice versa).
- Poor quantification.

Most formal science is expensive, often aimed at a more general level of question, integrates and tries to find truth over a wide spatial scale (synchronic strength) over short investigation spans (diachronic weakness). In contrast, for farmers limitations of academic experiments might be the appropriateness of the design or the applicability of the results to the site-specific conditions of a certain farm, and the lack of assessment under the complexity of annually changing farm conditions. Farmers' experiments are referenced against a long and culturally transmitted knowledge of how their local farm performed before the innovation was tried (diachronic strength) but may be less applicable to other farms, even ones nearby (synchronic weakness). Despite these weaknesses discussed in academia, for farmers their site specific experiments allows:

a) Local tuning of farmers' practice to the opportunities, threats and conditions of:
   - Ecology and biophysical features of the land and landscape (soil, climate, environmental history);

- Social needs and capacity for change (what works for the farmers' view and values);
- Economic resilience (financial capacity and equity, resilience to experiment and ride through failed experiments, financial drivers to improve a weak part of their economic performance);
- Governance-constraints (policy, regulations, view of their co-owners or sector co-operative).

b) Building resilience by increasing adaptability in a changing world – the keys to capturing new opportunities and counteracting new threats.

c) Immediate uptake—because the practitioners act as free agents to initiate and conduct the experiments, we know them to be relevant, of keen interest and likely to be immediately heeded by the main decision makers on the farmer—this removes the main barrier to external expert driven research actually being used.

For further research and theory building on learning and innovation in organic farming we propose to avoid replacing academic experiments with farmers' experiments, or putting higher values on farmers' experiments than on academic experiments. First, we call for an explicit appreciation of farmers' experiments and encourage further in-depth research on the details of the experimental process and the related intervening variables of farmers' experiments. Second, we want to encourage the inclusion of farmers' experiments in strategies for innovation in organic farming. Strategies could be on-farm research or participatory research as proposed by Moeskops and Cuoco (2014).

Nevertheless, intensity and kind of participation can vary significantly (Pretty, 1999). On-farm research might also be called as such when farmers simply provide land to academic scientists for academic experiments, and the degree of 'participation' might vary considerably. Both on-farm research as well as participatory research have the potential that the role of farmers remains quite passive and ignoring their experimental capacity.

Strategic research and innovation agendas for organic farming and food (Moeskops & Cuoco, 2014) must see organic farmers not only as actors providing the land for academic research, but farmers shall be included as co-researchers in various steps of the research process such as analyzing literature and empirical experiences, formulating research questions, developing the research design, monitoring, analysis and dissemination of results.

Here farmers can learn about aspects such as accuracy and validity of research designs, while scientists might benefit e.g. from a holistic research design and monitoring that can include factors beyond measuring controlled variables, and in doing so learning from farmers how to deal with complexity (comp. Hoffmann et al., 2007).

The usefulness of stakeholder participation in agricultural research was highlighted before and comprehensive participation frameworks were suggested to guide the participation process through self-reflection, informed discussion, and decision-making between project participants. These frameworks help to decide upon strengths and weaknesses of stakeholder inclusion in the steps of the research process and transcend common perceptions of the more participation the better (Neef & Neubert, 2011). Research done by Sewell et al. (2014) showed that farmers' learning can be highly promoted when farmers participate in a learning community with scientists and become part of a shared inquiry, because 'dialogue is not only a means of communication, but it is also a means to generate new ideas, negotiate understandings and build knowledge' (Sewell et al., 2014). Care has to be taken by the organic movement that standards and regulations encourage, but do not hamper farmers' experiments (Vogl et al., 2005).

Co-learning and co-production of knowledge (Akpo et al., 2014; Sewell et al., 2014) between organic farmers and academic scientists do have a yet underestimated and underused potential in opening the creative potential for innovations in organic farming. The potential might even increase by opening the scope from the farm perspective to a perspective on the whole supply chain of certain products or to a regional perspective and involving the stakeholders along the chain or in the region.

Organic farmers should not only be perceived as beneficiaries of innovations through cutting edge basic science or scientific experiments, or as hosts for on-farm experimentation, but also be explicitly supported in their capacity of being experimenters and perceived as genuine co-researchers. 'Farmers cannot resist tinkering with new techniques. They will do this whether outsiders tell them to do so or not, regardless of any project or agency's philosophy. Farmers are experimenters, no matter what happens, even if outsiders do not encourage them to do so' (Bentley et al., 2010).

We believe that more sustainable and resilient farming can emerge from better listening and integration of the practitioners' ways of knowing with the structured experiments of agronomists. Complementarity between farmers' and academics' experiments forms a strong partnership of approaches that collectively opens a wider choice set for farming practice options and local tuning. Together the most robust and lasting knowledge will emerge, but at the moment the two types of expert rarely communicate with each other.

REFERENCES

1. Akpo, E., Crane, T. A., Vissoh, P. V., & Tossou, R. C. (2014). Co-production of knowledge in multistakeholder processes: Analyzing joint experimentation as social learning. Journal of Agricultural Education and Extension. http://dx.doi.org/10.1080/1389224X.2014.939201
2. Bentley, J. W. (2006). Folk experiments. Agriculture and Human Values, 23(4), 451-462. http://dx.doi.org/10.1007/s10460-006-9017-1
3. Bentley, J. W., Van Mele, P., & Acheampong, G. K. (2010). Experimental by nature. Rice farmers in Ghana. Human Organization, 69(2), 120-137. http://dx.doi.org/10.17730/humo.69.2.r078vjvqx23675g1
4. Chambers, R. (1999). Rural Development. Putting the Last First, Essex, UK: Pearson Education Longman Ltd.
5. Chevalier, J. M., & Buckles, D. J. (2013). Participatory Action Research: Theory and Methods for Engaged Inquiry, London, UK: Routledge Chapman & Hall.
6. Critchley, W. R. S. (2000). Inquiry, initiative and inventiveness: Farmer innovators in East Africa. Physics and Chemistry of the Earth, 25(3), 285-288. http://dx.doi.org/10.1016/S1464-1909(00)00016-2
7. Critchley, W. R. S., & Mutunga, K. (2003). Local innovation in a global context: Documenting farmer initiatives in land husbandry through WOCAT. Land Degradation & Development, 14(1), 143-162. http://dx.doi.org/10.1002/ldr.537
8. Darnhofer, I. (2014a). Contributing to a transition to sustainability of agri-food systems: Potentials and pitfalls for organic farming. In S. Bellon & S. Penvern (Eds.), Organic farming, prototype for sustainable agricultures (pp. 439-452) Dordrecht, Germany: Springer. http://dx.doi.org/10.1007/978-94-007-7927-3_24
9. Darnhofer, I. (2014b). Resilience and why it matters for farm management. European Review of Agricultural Economics, 41(3), 461-484. http://dx.doi.org/10.1093/erae/jbu012
10. EU SCAR. (2012). Agricultural knowledge and innovation systems in transition – a reflection paper, European Commission, Brusseles, Belgium.
11. Folke, C., Carpenter, S. R., Walker, B., Scheffer, M., Chapin, T., & Rockström, J. (2010). Resilience thinking: Integrating resilience, adaptability and transformability. Ecology and Society, 15(4), 20-29.
12. Gerber, A., Hoffmann, V., & Kugler, M. (1996). Das Wissensystem im Ökologischen Landbau in Deutschland: Zur Entstehung und Weitergabe von Wissen im Diffusionsprozeß. Berichte über Landwirtschaft, 74(4), 591-627.
13. Hoffmann, V., Probst, K., & Christinck, A. (2007). Farmers and researchers: how can collaborative advantages be created in participatory research and technology development? Agriculture and Human Values, 24(3),355-368. http://dx.doi.org/10.1007/s10460-007-9072-2
14. Kummer, S, Milestad, R, Leitgeb, F., & Vogl, C. R. (2012). Building resilience through farmers' experiments in Organic Agriculture: Examples from Eastern Austria. Sustainable Agriculture Research, 1(2), 308-321. http://dx.doi.org/10.5539/sar.v1n2p308 www.ccsenet.org/sar Sustainable Agriculture Research Vol. 4, No. 3; 2015

15. Kummer, S., Leitgeb, F., & Vogl, C. R. (in prep.). Farmers' own research: The example of organic farmers' experiments in Austria. Under review. Journal for Renewable Agriculture and Food Systems.
16. Lawrence, D., Christodoulou, N., & Whish, J. (2007). Designing better on-farm research in Australia using a participatory workshop process. Field Crops Research, 104(1-3), 157-164. http://dx.doi.org/10.1016/j.fcr.2007.03.018
17. Leach, M. , Rockström, J., Raskin, P., Scoones, I., Stirling, A. C., Smith, A., ... Olsson, P. (2012). Transforming innovation for sustainability. Ecology and Society, 17(2), 11-17. http://dx.doi.org/10.5751/ES-04933-170211
18. Leitgeb, F, Kummer, S., Funes-Monzote, F. R., & Vogl, C. R. (2014). Farmers' experiments in Cuba. Renewable Agriculture and Food Systems, 29(1), 48-64. http://dx.doi.org/10.1017/S1742170512000336
19. Leitgeb, F., Funes-Monzote, F. R., Kummer, S., & Vogl, C. R. (2011). Contribution of farmers' experiments and innovations to Cuba's agricultural innovation system. Renewable Agriculture and Food Systems, 26(4), 354-367. http://dx.doi.org/10.1017/S1742170511000251
20. McIntyre, A. (2007). Participatory Action Research. Thousand Oaks, CA, USA: SAGE Publications.
21. McIntyre, B. D., Herren, H. R., Wakhungu, J., & Watson, R. T. (eds.). (2009). International Assessment of Agriculture Knowledge, Science and Technology for Development—IAASTD—Global Report, Washington, USA: Island Press.
22. Mittelstrass, J. (2011). On transdisciplinarity. Trames, 15(4), 329-338. http://dx.doi.org/10.3176/tr.2011.4.01
23. Moeskops, B., & Cuoco, E. (2014). Strategic Research and Innovation Agenda for Organic Food and Farming. Brussels, Belgium: TP Organics.
24. Moeskops, B., Blake, F., Tort, M-C., & Torremocha, E. (Eds.) (2014). Action Plan for Innovation and Learning, Brussels, Belgium: TP Organics.
25. Montgomery, D. C. (2009). Design and Analysis of Experiments (7 th edition). Hoboken, NJ, USA, John Wiley and Sons.
26. Neef, A., & Neubert, D. (2011). Stakeholder participation in agricultural research projects: A conceptual framework for reflection and decision-making. Agriculture and Human Values, 28(2), 179-194. http://dx.doi.org/10.1007/s10460-010-9272-z
27. Ninio, R., & Vogl, C. R. (2006). Organic farmers' experiments. Learning local knowledge. FWF project proposal. University of Natural Resources and Life Sciences, Vienna, Austria.
28. ODO—Oxford Dictionaries Online. (2010). Oxford Dictionaries Online. Oxford, UK: Oxford University Press.
29. Padel, S. (2001). Conversion to organic farming: A typical example of the diffusion of an innovation? Sociologia Ruralis, 40(1), 40-61. http://dx.doi.org/10.1111/1467-9523.00169
30. Pretty, J. N. (1991). Farmers' extension practice and technology adaptation: Agricultural revolution in 17th-19th century Britain. Agriculture and Human Values, 8(1-2), 132-148. http://dx.doi.org/10.1007/BF01579666
31. Pretty, J. N. (1995). Participatory learning for sustainable agriculture. World Development, 23(8), 1247-1263. http://dx.doi.org/10.1016/0305-750X(95)00046-F
32. Pretty, J. N. (1995). Regenerating Agriculture: Policies and Practice for Sustainability and Self-Reliance. London, UK: Earthscan.
33. Quiroz, C. (1999). Farmer experimentation in a Venezuelan Andean group.

34. Rajasekaran, B. (1999). Indigenous agricultural experimentation in home gardens of South India: Conserving biological diversity and achieving nutritional security. In G. Prain, S. Fujisaka, & M. D. Warren (Eds.), Biological and Cultural Diversity. The Role of Indigenous Agricultural Experimentation in Development (pp. 134-146). London. UK: Intermediate Technology Publications. http://dx.doi.org/10.3362/9781780444574.009
35. Reij, C., & Waters-Bayer, A. (2001). Farmer Innovation in Africa. A Source of Inspiration for Agricultural Development. London, UK: Earthscan.
36. Rhoades, R., & Bebbington, A. (1991). Farmers as experimenters. In B. Haverkort, J. van der Kamp & A. Waters-Bayer (Eds.), Joining Farmers' Experiments. Experiences in Participatory Technology Development (pp. 251-253), London, UK: Intermediate Technology Publications.
37. Rivera, W. M., Qamar, M. K., & Mwandemere, H. K. (2005). Enhancing Coordination Among AKIS/RD Actors: An Analytical and Comparative Review of Country Studies on Agricultural Knowledge and Information www.ccsenet.org/sar Sustainable Agriculture Research Vol. 4, No. 3; 2015 Systems for Rural Development (AKIS/RD). Rome, Italy: FAO.
38. Röling, N., & Brouwers, J. (1999). Living local knowledge for sustainable development. In G. Prain, S. Fujisaka & M. D. Warren (Eds.), Biological and Cultural Diversity. The Role of Indigenous Agricultural Experimentation in Development. (pp. 147-157). London. UK: Intermediate Technology Publications. http://dx.doi.org/10.3362/9781780444574.010
39. Saad, N. (2002). Farmer processes of experimentation and innovation. A review of the literature. CGIAR Systemwide Program on Participatory Research and Gender Analysis, Document number 21.
40. Sewell, A. M., Gray, D. I., Blair, H. T., Kemp, P. D., Kenyon, P. R., Morris, S. T., & Wood, B. A. (2014). Hatching new ideas about herb pastures: Learning together in a community of New Zealand farmers and agricultural scientists. Agricultural Systems, 125, 63-73. http://dx.doi.org/10.1016/j.agsy.2013.12.002
41. Smits, R. E., Kuhlmann, S., & Shapira, P. (2010). The theory and practice of Innovation Policy—An International Research Handbook, Williston, VT, USA: Edward Elgar Publishing. http://dx.doi.org/10.4337/9781849804424
42. Spielmann, D. J., & Birner, R. (2008). How innovative is your agriculture? Using innovation indicators and benchmarks to strengthen national agricultural innovation systems. Agriculture and Rural Development Diskussion Paper No. 41. The International Bank for Reconstruction and development / The World Bank, Washington, USA.
43. Stolzenbach, A. (1999). The indigenous concept of experimentation among Malian farmers. In G. Prain, S. Fujisaka & M. D. Warren (Eds.), Biological and Cultural Diversity. The Role of Indigenous Agricultural Experimentation in Development (pp. 163-171). London. UK: Intermediate Technology Publications. http://dx.doi.org/10.3362/9781780444574.012
44. Sturdy, J. D., Jewitt, G. P. W., & Lorentz, S. A. (2008). Building an understanding of water use innovation adoption processes through farmer-driven experimentation. Physics and Chemistry of the Earth, 33, 859-873. http://dx.doi.org/10.1016/j.pce.2008.06.022
45. Sumberg, J., & Okali, C. (1997). Farmers' Experiments: Creating Local Knowledge. London, UK: Lynne Rienner Publishers Inc.
46. Sumberg, J., Okali, C., & Reece, D. (2003). Agricultural research in the face of diversity, local knowledge and the participation imperative: Theoretical considerations. Agricultural Systems, 76(2), 739-753. http://dx.doi.org/10.1016/S0308-521X(02)00153-1

47. Tress, G., Tress, B., & Fry, G. (2005). Clarifying integrative research concepts in landscape ecology. Landscape Ecology, 20(4), 479-493. http://dx.doi.org/10.1007/s10980-004-3290-4
48. Tulloch, A. I. T., Joseph, L., Szabo, J. K., Martin, T., & Possingham, H. P. (2013). Realising the full potential of citizen science monitoring programs. Biological Conservation, 165, 128-138. http://dx.doi.org/10.1016/j.biocon.2013.05.025
49. van de Kerkhof, M., & Wieczorek, A. (2005). Learning and stakeholder participation in transition processes towards sustainability: Methodological considerations. Technological Forecasting and Social Change, 72(6), 733-747. http://dx.doi.org/10.1016/j.techfore.2004.10.002
50. Vogl, C. R., Kilcher, L., & Schmidt, H.-P. (2005). Standards and regulations of organic farming: Moving away from small farmers' knowledge? Journal for Sustainable Agriculture, 26(1), 5-26. http://dx.doi.org/10.1300/J064v26n01_03
51. Warren, M. D. (Eds.). Biological and Cultural Diversity. The Role of Indigenous Agricultural Experimentation in Development. London, UK: Intermediate Technology Publications.
52. Wortmann, C. S., Christiansen, A. P., Glewen, K. L., Hejny, T. A., Mulliken, J., Peterson, J. M., ... Zoubek, G. L. (2005). Farmer research: Conventional experiences and guidelines for alternative agriculture and multi-functional agro-ecosystems. Renewable Agriculture and Food Systems, 20, 243-251. http://dx.doi.org/10.1079/RAF2005110.

CHAPTER 7

Supporting Innovation in Organic Agriculture: A European Perspective Using Experience from the SOLID Project

Susanne Padel, Mette Vaarst, and Konstantinos Zaralis

7.1 INTRODUCTION

Innovation and agriculture have always gone 'hand-in-hand' because working with dynamic geographic, climatic, market and political conditions requires constant change (EC-SCAR, 2012). According to Hoffman et al. (2007) farmers have been developing agricultural practices since the beginning of agriculture, about 10,000 years ago.

Their innovative power can be seen in many crops species grown and in different animal breeds, in the development of new production systems, farm machinery and equipment and also in social innovations (Hoffmann et al., 2007). Today innovation is seen as the primary instrument for overcoming the sustainability challenges of agriculture at the beginning of 21st century, such as food security, climate change and the conservation of natural resources. The European

---

Innovation Partnership for Agricultural Productivity and Sustainability (EIP-AGRI) was set up in response to these challenges (EIP-AGRI, 2012).

Organic farming is recognized as one source for innovation helping agriculture to overcome such challenges: "Organic farming with its stringent rules on external input use has to be even more innovative to solve production problems, sometimes opening up new avenues" (McIntyre et al., 2009, p. 384). The European Technology Platform TP Organics describes organic farms as "creative living laboratories for smart and green
innovations" (Padel et al., 2010). Organic farming can make an important contribution and will continue to innovate in order to adapt to changing conditions in the climate as well as in the developing market.

However, innovation in agriculture is currently frequently understood as referring exclusively to the need for new inputs and technologies that originate from research (Röling, 2009). Garnet and Godfray (2012) referred to this as technological optimism in the debate about sustainable intensification. Much of the agricultural research effort in the last century has been concerned with developing and using external inputs (such as fertilizers and germplasm). Understanding how farmers adopt such science derived innovation was the starting point for the model of adoption and diffusion of innovation (Rogers, 1983). This led to the technology transfer model of the green revolution, where research was seen as the main generator of innovation that had to be transferred to and adopted by the farmers. The adoption/diffusion model was applied to organic farming by Padel (2001). She concluded that early organic farmers share many characteristics with other innovators. However, organic farming could not be characterized as a typical innovation, because it requires complex change, brings often no recognized economic advantage, conflicts with some rural values and is knowledge-intensive, whilst access to information is limited (Padel, 2001). This clearly limits the usefulness of the adoption/diffusion model to understand innovation in the organic sector. In Europe, the conceptual framework of innovation systems is gaining in importance for agriculture (EC-SCAR, 2011) and is underlying the new instrument of the Innovation Partnership of the European Union (EIP-AGRI, 2012).

In this paper we explore how innovation occurs within the organic sector in Europe and how this process can be further supported, using framework of innovation systems and experiences from the ongoing project 'Sustainable Organic Low-Input Dairying (SOLID)'. We first describe the approach of encouraging stakeholder-led innovation that was used in the project and present experiences gained so far. Based on selected examples, we discuss how innovation potentially can support sustainable development within the farming sector. This challenges the widespread perception that innovation in agriculture is mainly about new technologies and inputs and illustrates the importance of using active sharing of existing knowledge and of close collaboration between farmers and researchers in supporting innovation in this sector.

7.1.1 FROM TECHNOLOGY TRANSFER TO SUPPORTING THE INNOVATION SYSTEM

Innovation is a broad concept defined as the development, introduction and application of a new or significantly improved product (good or service), a new marketing method or a new organizational method in business practice, workplace organization or external relations where an economic or social benefit is assumed for individuals, groups or entire organizations (OECD/Eurostat 2005). The concept of 'innovation' is not restricted to invention or a new idea itself, but includes also the embedding of an idea in the relevant sector (Schumpeter et al., 1980).

However, within agriculture innovation is seen mainly as the search for new inputs and technologies (Röling, 2009) while the potential of social/societal innovation for achieving societal and political goals is not recognized (Bokelmann et al., 2012). This maybe not so surprising, given the long period during which "efficiency came ... to mean the application of the new agricultural technologies, which were beginning to emerge onto the market" (Morgan & Murdoch, 2000). In arable production, the farmers' 'know-how' was replaced by 'know-what', i.e. what input to use and when (Morgan & Murdoch, 2000). This 'technical optimism' remains strong in contemporary thinking about sustainable intensification of agriculture in the UK, but the need for new perspectives is beginning to be recognized (Garnett & Godfray, 2012).

In contrast, the concept of innovation systems describes innovation as an interactive evolutionary process, from invention to successful adoption by the target group with different participants involved at various stages (Smiths et al., 2010). Innovation occurs when networks of organizations come together with the institutions and policies that affect innovative behavior and bring new products and processes into economic and social use (various authors cited by Hall et al., 2005). Innovation becomes an emergent property not only of science or the market, but of interaction between stakeholders that allows opportunities to develop (Röling, 2009). The relevance of this concept for agriculture in Europe is increasingly recognized (e.g. Bokelmann et al., 2012, EC SCAR, 2012). The concept of innovation systems differs from the technology transfer framework also in the types of innovation considered, with the former focuses mainly new technologies, whereas the later differentiates between consumer driven, technology driven and organizationally driven pathways to innovation. The European Innovation Platform for Agricultural Productivity and Sustainability that wants to use partnerships and bottom-up approaches, linking farmers, advisors, researchers, businesses, and other participants in so called Operational Groups is based on this concept (EIP-AGRI, 2012).

Following on from Farmer First (Chambers et al., 1989), many authors argue that it is important to put the farmer back at the center of knowledge production (e.g. MacMillan and Benton, 2014). Farmer involvement is thereby critical in all stages of the process, so that novel technologies and practices can be learned directly and then adapted to particular agro-ecological, social and economic circumstances (Pretty et al., 2011). Others refer to 'co-innovation' that can involve a diverse range of participants other than farmers, such as rural entrepreneurs, regional governments, researchers and knowledge brokers (EC-SCAR, 2011; Knickel et al., 2009).

7.1.2 THE ROLE OF KNOWLEDGE IN INNOVATION IN THE ORGANIC AGRICULTURE SECTOR

Innovation is the application of knowledge to achieve desired social and/or economic outcomes. This knowledge may be acquired through learning, research or experience, but the process is not considered as innovation until the knowledge is applied more widely (Hall et al., 2005). Sustainable agriculture makes productive use of human and social capital in the form of knowledge sharing to adapt and innovate to resolve common landscape-scale problems (Pretty et al., 2011). The techniques and practises used in organic farming are knowledge intensive (Lockeretz, 1991) and knowledge sharing between farmers is at the heart of the agrecology movement (Wezel et al., 2009).

Faced with new challenges of productivity, environmental change, and market conditions, organic farmers also have to evolve and innovate. Some innovation in organic farming occurs through the reapplication of existing knowledge. The European Technology Platform TP Organics referred to 'know-how innovation' to distinguish innovation that relies entirely on recombining and applying existing knowledge from other technological or social/societal and organizational innovation (Padel et al., 2010). Examples of such 'know-how innovation' include securing essential supply of vitamins and minerals in animal diets from natural sources, using composts for plant protection or encouraging predators by creating suitable habitats (e.g. flowering field margins). The definition of 'know-how innovation' used by the platform is very similar to the concept of exploitative knowledge strategies as compared to explorative ones (Li et al., 2008; March, 1991). In an exploitative strategy firms focus on levering existing knowledge to rapidly create new organizational products and processes, whereas in an explorative one they strive to develop capabilities to create or acquire new knowledge. Knowledge exploitation fits well into innovation systems concepts, whereas the explorative knowledge strategy has similarities to concept of 'technological innovation' (e.g. new germplasm or new machinery). TP organics argued that 'know-how innovation' is crucial to the organic farmer's ability to innovate, i.e. to respond effectively to new challenges, such as saving and protection of natural resources, and for improving the multi-functionality and sustainability of agriculture (Padel et al., 2010).

7.2 APPROACH TO ENCOURAGE INNOVATION THROUGH STAKEHOLDER ENGAGEMENT AND PARTICIPATORY RESEARCH IN THE SOLID PROJECT

The European Union (EU) funded SOLID project (Sustainable Organic Low-Input Dairying) carries out research to improve the sustainability of low-input/organic dairy systems, aiming to improve the health and welfare, productivity and product quality by better understanding how contrasting genotypes adapt to such conditions, and to improve the supply of nutrients from forages and by-products through the use of novel feeds. The five-year project also performs environmental, economic and supply chain assessments and promotes knowledge exchange. We report here from one work package that aims to facilitate innovation by actively involving farming stakeholders (i.e. organic and low-input dairy farmers, farmer groups and farm advisors) and stakeholder partners together with researchers in a participatory approach.

We used a farmer-led approach to identify problems of organic and low-input dairy farmers and develop and evaluate some potentially innovative solutions. In addition to research partners (from institutes and universities), the project also involved enterprise partners (small and medium size milk companies (SMEs) that work with groups of organic and low input dairy cow and goat producers in nine countries. The participatory approach progressed in four steps.

1. Identifying topics where farmer feel knowledge or innovation is needed.
2. Developing appropriate research approaches and experimental procedures to test innovative solutions for topics identified in Step 1.
3. Carry out the proposed research with small number of farms or groups of farmers (between one and five per country).
4. Report on the lessons learned and communicate the result to farmers, consultants and researchers.

The work is still on-going so experience has so far mainly been gained with the first and the second step of this approach which are described in some detail here.

7.2.1 IDENTIFICATION OF POTENTIAL TOPICS FOR PARTICIPATORY RESEARCH

The emphasis in this step was on working with producers to identify topics for the development, implementation and analysis of relevant, producer-led projects. At first, we carried out a rapid sustainability assessment on ten farms in each country, encouraging the farmers to think not only about immediate practical needs but reflect on the overall sustainability of their farms. Farms were chosen among the SME members to illustrate the range within low-input and organic farms in terms of size, intensity/level of input use, breeds, products, marketing channels and geographical area in the respective country/region and to highlight potential sustainability hotspots.

The assessment of different strands of sustainability used a tool developed by Organic Research Centre adapted to the project (Gerrard et al., 2011; Marchand et al., 2014). After some initial hesitation, both farmers and researchers viewed the process mainly positively, but expressed also questions about specific data requests and the validity of some indicators.

The results of the sustainability assessments were presented at meetings, attended by between 10 and 25 farmers, aimed at identifying research needs and constraints of the industry and to formulate potential solutions which could subsequently be tested. A common protocol for the workshops encouraging farmers to discuss successes and innovative or unusual practices on their farms provided a link between everyday practical issues and sustainability, before moving to ideas how to further develop strength and address the perceived problems. The facilitators' role was to draw out areas of common interest related to the farmers' practical situations as well as remaining relevant to the overall issue of sustainability (see Leach et al., 2013 for details of approach).

7.2.2 DEVELOPING THE APPROPRIATE RESEARCH METHOD

Further discussion between the farmers, SMEs and researchers lead to the narrowing down of suitable research topics and to the setting-up of specific on-farm research projects. Not all topics and themes initially suggested could be investigated, because only a limited number of studies could be carried out. The following methods were used:

- Farm case studies were based on monitoring certain aspects on a single farm, using a variety of data collection methods both quantitative and qualitative. This allows for observations to be made in context of a specific farm (see Maxwell, 1986; Padel, 2002). In some cases we used comparative case studies, where this approach was extended to several farms and observations could be compared between different farms.
- On-farm trials introduced a specific treatment (e.g. use of new feed resources) which was compared with a control group or with performance before the treatment was introduced.

Several projects were carried out as group discussion, which are the facilitated exchange of farmer experience and other knowledge sources among participating farmers with the aim to improve practice. This approach is inspired by the Farmer Field Schools (SUSTAINET EA, 2010), the Danish concept of Stable Schools (Vaarst et al., 2007), the approach of field labs developed in the UK (MacMillan & Benton, 2014) and focus groups.

The choice of method depended on the topic under study and in some cases involved the combination of some of the elements. A common template for reporting outlining also the farmers' background to a specific topic and the experience with the approach was developed.

7.3 EXPERIENCE SO FAR

7.3.1 IDENTIFYING RESEARCH AND INNOVATION NEEDS

Evaluating the sustainability of selected farms was intended to 'set the scene' and consider sustainability in its broadest sense whilst identifying suitable topics for participatory research. The results illustrate the diversity of low-input and organic dairy farms in the nine countries in terms of size and intensity. Cow farms varied from less than 20 ha (Austria and Italy) to more than 400 ha (Denmark, UK), with herd sizes ranging from nine (Finland) to over 300 cows (Italy, Denmark, UK) and milk yields ranging from less than 2500 kg/cow (Austria and Romania) to more than 8000 l/cow (Denmark). There was landless dairy goat farming in Spain and Flanders, but also grazing on more than 300 ha of common land in Spain and Greece with herd sizes between 22 goats (Spain) and 1150 (Belgium) and milk yields between 117 and 900 l/year. After the assessment, twelve workshops were held to identify knowledge and research needs from the farmers' point of view. They were attended by 161 dairy producers (the majority of which kept cows) in nine countries, and by some staff of the SMEs and facilitated by researchers and/or consultants. The farmers welcomed the opportunity to participate, related to their view that research specifically providing knowledge for organic/low input production was lacking. Further details of the outcomes of the sustainability and the workshops are reported by Leach et al. (2013).

Carrying out a structured sustainability assessment stimulated discussion, both during the visit and in the group meetings. Most farmers' own perception of sustainability included economic sustainability. Exposed to changing markets they do not see any future in farming, if they cannot run the businesses profitably, but the farmers were also aware of some other components of sustainability. The use of the tool encouraged them to think about the wider aspects. Some topics initially viewed sceptically, sparked interest and led to further discussion and some topics emerged from the sustainability assessments. For example, biodiversity management was discussed at first very critically among the mountain farmers in Austria but was eventually chosen as the research topic. Farmers in Denmark and in the UK strongly felt that they should improve in relation to greenhouse gas emissions by using more renewable energy and to diversify their farms.

Topics for which the farmers wanted to see further research effort have been summarized under the broad headings of animal feeding and forage production, natural resource management, animal management, product differentiation and marketing.

7.3.2 FEEDING PRACTICES AND FORAGE PRODUCTION

Topics included forage quality (i.e. protein), forage productivity and reliability, establishment and utilization of forage crop (such as diverse swards) and cultivation and feed value of protein rich crops (such as lupins, beans, and lucerne). Many dairy farmers reported not feeling confident about growing these crops, despite existing information on the subject. There were also a range of very specific suggestions, such as equipment and energy needs for drying forage (Austria), using various plant species (including for browsing) and identifying drought resistant plants and varieties (Italy, Romania, Spain, UK). Interest in diverse pastures was related to several different expected benefits, such as using them as natural sources for the supply of minerals (mainly in Denmark), improvements in forage quality (UK), creating marketing opportunities through improved product qualities (Austria, Italy) and improving soil quality (UK, see 3.2.2). The Greek farmers were interested in the use of irrigation for pastures. The use of novel forage is also investigated in other parts of the SOLID project (e.g. Rinne et al., 2014).

Some unusual feeding practices used on farms could be applied more widely, illustrating the potential value of knowledge sharing. Goat farmers in the Netherlands used by-products from a muesli factory and Austrian cow farmers 'grass cobs' to reduce purchased concentrate. The cultivation of some vetches as feedstuffs for goats was commonly place in some countries, but considered innovative elsewhere. Romanian farmers referred to trying 'forgotten' feeds such as turnips, millet and sorghum. The discussions and suggestions for further research show that good use of forage is of vital importance for low-input and organic dairy farms, but there is a lack of confidence in the reliability of forage production both in terms of quantity and quality.

7.3.3 NATURAL RESOURCE MANAGEMENT

Farmers in the UK wanted a better understanding of the soil to be able to diagnose potential problems with declining productivity under organic conditions and suggested research into topics of increasing soil organic matter. Austrian farmers discussing manure application were not fully aware of the considerable amount of information that already exists on this subject.

Farmers in Denmark and Finland showed the greatest concern about energy use and climate change, perhaps as a result of national policies and legislation and the demonstration of energy saving practices. The Austrian farmers used biomass from their own forest to fuel a hay drying installation. The assessment of environmental impact of low-input and organic dairy farming is a topic that is also covered elsewhere in the SOLID Project (e.g. Hietala et al., 2014).

7.3.4 ANIMAL MANAGEMENT

Despite some ongoing research on the subject, the choice of cow breeds and animals best suited to low input and/or organic systems was raised as research need in Denmark, Austria, Italy and the UK and by the goat farmers in Greece. The suitability of breeds for organic and low-input systems is also investigated in on-station experiments of the SOLID project (e.g. Horn et al., 2013).

Although animal health and welfare scored well in sustainability assessment, the farmers identified at least one health or welfare related issue in each workshop, including using fewer antibiotics (UK), improving health and longevity (Finland), parasite and disease control in goats (Belgium) and determining risk factors for neonatal losses and sub-clinical mastitis (Greece), even if on many of these subjects, research knowledge is available.

Less common practices with innovative potential included seasonal calving and rearing calves on mothers and nurse cows (UK and Denmark), once-a-day milking (UK) and extending goat lactations (Belgium).

7.3.5 PRODUCT DIFFERENTIATION AND MARKETING

Farmers were interested in product differentiation and in improved communication with consumers about the value of their products. One Italian farm aimed to standardize a high forage diet to market milk with a high nutritional value. This topic has been studied in several research projects (several authors cited by Leach et al., 2013) but so far farmers or SMEs have not developed related differentiation strategies. The topic was not taken up further in this project.

Farmers used specific attribute in selling directly to the public, e.g. in connections with agro-tourism in Austria, by offering a good product range in Greece or by selling raw milk through authorized dispensing machines in Romania. In Spain, one cheese-making farm developed an 'a la carte' strategy, targeting high-end restaurants for different types of flavored goat cheeses (matured in olive oil, with herbs).

7.3.6 SETTING UP THE PARTICIPATORY RESEARCH PROJECTS

The next stage involved further discussions to narrow down the topics, because the number of projects in each country was limited. Setting the 'right' research question is important for the successful conduction and the quality of any research and this is equally important for participatory studies. In this case, the experience of the farmers' in what treatments can be implemented and what indicators can be monitored under practical conditions had to be brought together with the researchers' knowledge of experimental design, data analysis and statistics. The process is illustrated with two UK examples.

One UK farmer, with the aim of increasing soil organic matter, established very diverse and herb-rich swards and grazes in an extended rotation, along the lines of "mob grazing". The topic was of interest also to several other UK farmers, so a case study for monitoring the farm was developed (Leach et al., 2014).

The UK SME partner wanted to further explore the link between diet and cow health on a number of farms. However, given the variability on management practices across farms and the difficulty in identifying parameters that could be manipulated under practical conditions of different farms made clear that this question was not suitable for this type of research. As a result we opted for an approach that can account for potential confounding effects due to different farm practices and conditions with the aim to study how different farm management practices can affect the concentration of iodine in milk in view of the iodine supplied by the feed which was also of great interest to the SME partner.

The final choice of topics summarized in Table 7.1 reflects priority for the farmers and suitability for on- farm research, and a suitable approach was developed using the different methods described in Section 7.2.2. Although farmers in several countries were also very interested in product differentiation and marketing no on-farm experiments were selected in this area, but the results of Austrian, Italy studies could support this in future.

TABLE 7.1 Topics of farmer-led research in the SOLID* project and the adopted study methods

Thematic area	Topic	Approach	Country
Feeding and forage	Home grown proteins	On-farm trials	Finland
	Use of by-products	On-farm trials	Spain, Romania
	Irrigation of pasture	On-farm trial	Greece
Natural resources use and environmental impact	Soil management, pasture productivity and grazing	Farm case study with monitoring of forage production	United Kingdom
	Responding to climate change	Moderated discussion group and farm case studies	Denmark
	Impact of different protein sources on carbon footprint	Case study using LCA (Life Cycle Analysis) method	Italy
	Impact of intensification on biodiversity	Comparative farm case studies with assessments and modelling	Austria
Animal management	Reducing antibiotic use	Moderated discussion group followed by on-farm trials	United Kingdom (jointly with DFF~)
	Herbs in pasture	Comparative case studies	Denmark
	Maternal/nurse cow rearing of calves	Farm case study with monitoring of calf growth	United Kingdom and Denmark
	Impact of farm practices on concentration of iodine in milk	Comparative farm case studies	United Kingdom

Source: Own data
*for a description of the protocols and future publication of results please see www.solidairy.eu, SOLID (Sustainable Organic Low-Input Dairying)

~DFF is the Duchy Future Farming program of the Soil Association (http://www.soilassociation.org/fieldlabs).

7.4 DISCUSSION

7.4.1 INNOVATION IN ORGANIC AGRICULTURE THROUGH KNOWLEDGE EXPLOITATION

The research and innovations topics discuss by the SOLID farmers include many examples of exploitative innovation strategy (see March 1991) where already a considerable amount of research exists. For example, sustainability could be improved by mobilizing knowledge about growing and feeding many different forage crops. Incremental change based on better exploitation of existing knowledge by producers, e.g. through re-combining it in different ways, appears very important for further development of organic and low-input farms, but is not likely to be restricted to these sectors. However, examples of breeding new varieties of forage legumes or other feed crops with high protein content illustrate that there also is a need for explorative innovation.

This importance of both explorative and exploitative strategies is also reflected in responses of low-input and organic dairy farmers to a list of innovation statements that they were shown in another part of the SOLID project. The aim of the survey was to contrast views about acceptability of innovation statements between different actors in the supply chain (farmers, processors/retailers and consumers) in Belgium, Finland, Italy and the UK, using Q sort methodology (Nicholas et al., 2014). The farmers strongly liked statements referring to exploitative innovation, such as developing techniques to improve feed and forage quality, reduce the use of purchased concentrate as well as improving feed quality and efficiency and animal welfare. They disliked some explorative statement that referred to what they saw as 'unnatural' innovation, such originating from GM or semen sexing, but strongly liked statements of 'developing of new forage varieties specific for low input and organic farming' (Nicholas et al., 2014).

7.4.2 THE ROLE OF OPEN-ACCESS IN SUPPORTING THE INNOVATION PROCESS FOR SUSTAINABLE DEVELOPMENT

In our view, it is also necessary to reflect on who will benefit from future innovation in organic agriculture or related systems. Some innovation will generate specific benefits for farmers, such as increased profitability, but much will generate public benefits, such as reduced natural resource use, improvement of soil fertility, of biodiversity and of animal health. Such innovation is a necessary part of sustainable development. We agree with the conclusion of Buckwell et al. (2014) that as part of sustainable intensification of European agriculture the 'knowledge per hectare' should to be intensified, including knowledge about how to manage the ecosystem services on which agriculture relies. We would like to emphasize again that 'innovation' is not necessarily a product, but a reflected part of continuous process, which involves creative thinking and knowledge sharing through learning in communities. In the United States, the idea of the open-access knowhow to farming is well established but also in Europe there are some good examples of open-access, for example the research archive for organic agriculture (e.g. http://orgprints.org).

7.4.3 HOW CAN LOCALLY GENERATED KNOWLEDGE BE VALUABLE IN OTHER CONTEXTS?

We believe there are three main reasons that limit the universal nature of locally-generated knowledge: ecology, economic and market context, and social/cultural values. Knowledge about the ecology of the given environment is location-specific and becomes only transferable where workable model of the ecological interactions under various pedo-climatic conditions exists. The interest in increasing home-grown protein crops illustrates that sharing relevant knowledge about specific crops could help the farmers to become more confident in growing them, but uncertainty remains under which conditions which corps are worth trying. And organic sector development will influence access to specific organic inputs, for example for organic feed. Finally, existing knowledge is also specific to personal goals and styles, social norms and cultural contexts. Curry and Kirwan (2014) conclude that the complex set of objectives, values and styles of implementing sustainability agriculture at various locations has an impact on how much knowledge can be seen as universal.

Farmers are aware that research often excludes variables that they know to be important for their decision-making but they may feel unable to express these clearly. This is likely to be a reason why they often have greater trust of farmers than of other experts. The farmer (tacit) knowledge, grounded in the farmers' observations of the various parts of their system and of the local environment, is important for the success or failure of new practices. Therefore farmers need to be recognized as active contributors to generating innovation rather as than passive recipients of knowledge transfer.

However, science derived knowledge cannot be replaced by context or farmer knowledge. Science derived knowledge needs to include basic ecological principles and the state of resources and ecosystem services on which agriculture depends. To foster innovation, this scientific knowledge must be complemented by location specific knowledge related to the ecology, economics and culture. And farmers and researchers as the two main actor groups contributing knowledge (as well as advisors, consultants and other intermediaries) need work closely together. The experience in SOLID has shown that farmer-led research is a good way to stimulate this dialogue between the farmers and scientists as equal partners in trying to find solutions to the problems experienced by the farmers and develop sustainability.

7.4.4 SUPPORTING ACTIVE FARMER LEARNING FOR INNOVATION FOR SUSTAINABILITY

How should such knowledge exchange systems be organized to support innovation for sustainability of agriculture? This shift away from dissemination and 'technology transfer' towards recognizing the role of farmers implies learning as active knowledge construction (Koutsouris, 2012). Farmers need to become confident observers of their own systems, so that they can learn the lessons, draw their own conclusions and recombine elements to develop their own solutions. The discussions among UK farmers identifying soil fertility as a research topic illustrate this point: some organic farmers had observed that productivity of some of their swards had dropped, but did not feel that could identify the causes and implement solutions using standard soil analysis so they wanted to know more about biological soil processes. Ongoing activities in the project are aimed at testing simple diagnostic tools that the farmers can use.

A study of learning and Innovation Networks for Sustainable Agriculture (LINSA) concluded that such groups' need to adopt a strong focus on the process of learning to effectively support innovation in the farming sector. In particular, the dimension of social learning with groups of farmers has received attention, but this is not to say that education in schools and colleges does not deserve to be considered to foster change. In the LINSA groups, social learning emerges from a shared interest in a problem, challenge or activity and all the actors bring all their expertise to the table. Social learning is linked to processes of trust building, trial and error and of mutual support and can provide answers to very complex problems, because mutual reflection on knowledge and consciously hearing different perspectives on one common issue will enhance the portfolio of potential solutions (Moschitz et al., 2014).

7.5 CONCLUSIONS

The conceptual framework of innovation systems uses a broad definition of innovation and describes it as the outcome of a stakeholder interaction process. This framework is more suited to understand and support innovation for sustainability and within organic agriculture than the technology transfer model.

Farmers are active contributors to agricultural innovation, who contribute context specific knowledge as well as their creativity. The restrictions of certain inputs and the focus and direction of organic standards encourage organic farmers to try a range of alternative solutions.

Knowledge exploitative and explorative innovation strategies are likely to be equally important to improve sustainability of organic and low-input dairy farming. An example of exploitation is improving forage production and utilization, and examples of explorative innovation are new forage cultivars and species.

Innovation for sustainability generates private but also much public benefits, such as reduced natural resource use, improvement of soil fertility, of biodiversity and of animal health. The open-access model of knowledge sharing is compatible with supporting this process and should be more widely used.

Knowledge exchange supporting innovation for sustainability needs to bring science-based and farmer (tacit) knowledge together. Farmer-led research is an effective way for researchers and the farmers together to develop sustainability of agriculture.

REFERENCES

1. Bokelmann, W., Doernberg, A., Schwerdtner, W., Kuntosch, A., Busse, M., König, B., ... Stahlecker, T. (2012). Sektorstudie zur Untersuchung des Innovationssystems der deutschen Landwirtschaft. Berlin: Humboldt-Universität.
2. Buckwell, A., Nordang Uhre, A., Williams, A., Polakova, J., Blum, W. E. H., Schiefer, J., ... Haber, W. (2014). The sustainable intensification of European agriculture. Brussels: RISE Foundation, www.ccsenet.org/sar.
3. Chambers, R., Pacey, A., & Thrupp, L. A. (Eds.) (1989). Farmer First: Farmer Innovation and Agricultural Research, London: Intermediate Technology Publications. http://dx.doi.org/10.3362/9781780440149

4. Curry, N., & Kirwan, J. (2014). The Role of Tacit Knowledge in Developing Networks for Sustainable Agriculture. Sociologia Ruralis, 54 (3), 341-361. http://dx.doi.org/10.1111/soru.12048
5. EC-SCAR. (2011). The 3rd SCAR Foresight Exercise . Standing Committee on Agricultural Research (SCAR). Brussels: European Commission.
6. EC-SCAR. (2012). Agricultural Knowledge and Innovation Systems in Transition—A reflection paper. Standing Commitee on Agricultural Research - Collaborative Working Group on Agricultural Knowledge and Innovation System (CWG AKIS). Brussels: European Commission.
7. EIP-AGRI. (2012). European Innovation Partnership Agricultural Productivity and Sustainability (EIP AGRI). Brussels: European Commission.
8. Garnett, T., & Godfray, H. C. J. (2012). Sustainable intensification in agriculture. Food Climate Research Network and the Oxfrod Martin Programme on the Future of Food. Oxford: Univeristy of Oxford.
9. Gerrard, C. L., Smith, L. G., Padel, S., Pearce, B., Hitchings, R., Measures, M., & Cooper, N. (2011). OCIS public goods tool development. Research report. Newbury: Organic Research Centre.
10. Hall, A., Mytelka, L., & Oyeyinka, B. (2005). From Innovation systems: Implications for agricultural policy and practice - CGIAR- ILAC Source-book Chapter 3. Retrieved from
11. http://www.cgiar-ilac.org/content/chapter-3-innovation-systems
12. Hietala, S., Smith, L., Knudsen, M., Kurppa, S., Padel, S., & Hermansen, J. (2014). Carbon footprints of organic dairying in six European countries—real farm data analysis. Organic Agriculture, 1-10.
13. Hoffmann, V., Probst, K., & Christinck, A. (2007). Farmers and researchers: How can collaborative advantages be created in participatory research and technology development? Agriculture and Human Values, 24, 355-368. http://dx.doi.org/10.1007/s10460-007-9072-2
14. Horn, M., Steinwidder, A., Gasteiner, J., Podstatzky, L., Haiger, A., & Zollitsch, W. (2013). Suitability of different dairy cow types for an Alpine organic and low-input milk production system. Livestock Science, 153, 135-146. http://dx.doi.org/10.1016/j.livsci.2013.01.011
15. Knickel, K., Tisenkopfs, T., & Peter, S. (Eds.). (2009). Innovation processes in agriculture and rural development: Results of a cross-national analysis of the situation in seven countries, research gaps and recommendations, Frankfurt (Main), Germany: IfLS Frankfurt at Goethe Universit.
16. Koutsouris, A. (2012). Facilitating Agricultural Innovation Systems: a critical realist approach. Studies in Agricultural Economics, 114, 64-70. http://dx.doi.org/10.7896/j.1210
17. Leach, K., Gerrard, C. L., & Padel, S. (Eds.). (2013). Rapid sustainability assessment of organic and low-input farming across Europe and identification of research needs, Hamstead Marshall, Newbury: Organic Research Centre.
18. Leach, K., Palomo, G., Waterfield, W., Zaralis, K., & Padel, S. (2014). Diverse swards and mob grazing for dairy farm productivity: A UK case study. In Thunen Report, no. 20. Braunschweig: Thunen-Institut, pp.1155-1158.
19. Li, Y., Vanhaverbeke, W., & Schoenmakers, W. (2008). Exploration and exploitation in innovation: Reframing the interpretation. Creativity and innovation management, 17 (2), 107-126. http://dx.doi.org/10.1111/j.1467-8691.2008.00477.x

20. Lockeretz, W. (1991). Information requirements of reduced chemical production methods. American Journal of Alternative Agriculture, 6 (2), 97-103. http://dx.doi.org/10.1017/S0889189300003957
21. MacMillan, T., & Benton, T. G. (2014). Agriculture: Engage farmers in research. Nature, 508 , 25-27. http://dx.doi.org/10.1038/509025a
22. March, J. G. (1991). Exploration and exploitation in organizational learning. Organization science, 2 (1), 71-87. http://dx.doi.org/10.1287/orsc.2.1.71
23. Marchand, F., Debruyne, L., Triste, L., Gerrard, C., Padel, S., & Lauwers, L. (2014). Key characteristics for tool choice in indicator-based sustainability assessment at farm level. Ecology and Society, 19 (3). http://dx.doi.org/10.5751/ES-06876-190346
24. Maxwell, S. (1986). The role of case studies in farm systems research. Agricultural Administration, 21, 147-180. www.ccsenet.org/sar Sustainable Agriculture Research Vol. 4, No. 3; 2015 http://dx.doi.org/10.1016/0309-586X(86)90083-X
25. McIntyre, B. D., Herren, H. R., Wakhungu, J., & Watson, R. T. (Eds.). (2009). Agriculture at a crossroads [Global Report by the International Assessment of Agricultural Knowledge, Science and Technology for Development (IAASTD)]: Synthesis Report, Washington DC: International assessment of agricultural knowledge, science and technology for development (IAASTD).
26. Morgan, K., & Murdoch, J. (2000). Organic vs. conventional agriculture: knowledge, power and innovation in the food chain. Geoforum, 31 (2), 159-173. http://dx.doi.org/10.1016/S0016-7185(99)00029-9
27. Moschitz, H., Tisenkopfs, T., Brunori, G., Home, R., Kunda, I., & Sumane, S. (2014). Final report of the Solinsa project. Frick: FIBL.
28. Nicholas, P. K., Mandolesi, S., Naspetti, S., & Zanoli, R. (2014) Innovations in low input and organic dairy supply chains—What is acceptable in Europe? Journal of Dairy Science, 97 (2), 1157-1167. http://dx.doi.org/10.3168/jds.2013-7314
29. OECD/Eurostat. (2005). Oslo Manual. Guidelines for Collecting and Interpreting Innovation Data, (3rd Edition). Paris: OECD Publishing.
30. Padel, S. (2001). Conversion to organic farming: A typical example of the diffusion of an innovation? Sociologia Ruralis, 41 (1), 40-61. http://dx.doi.org/10.1111/1467-9523.00169
31. Padel, S. (2002). Conversion to organic milk production: the change process and farmers information needs. PhD Thesis, Aberystwyth: University of Wales.
32. Padel, S., Niggli, U., Pearce, B., Schlüter, M., Schmid, O., Cuoco, E., ... Micheloni, C. (2010). Implementation Action Plan for organic food and farming research. Brussels: TP Organics. IFOAM- EU Group.
33. Pretty, J., Toulmin, C., & Williams, S. (2011). Sustainable intensification in African agriculture. International Journal of Agricultural Sustainability, 9 (1), 5-24. http://dx.doi.org/10.3763/ijas.2010.0583
34. Rinne, M., Dragomir, C., Kuoppala, K., Smith, J., & Yanez Ruiz, D. (2014). Novel feeds for organic dairy chains. Organic Agriculture, 4 , 275-284. http://dx.doi.org/10.1007/s13165-014-0081-3
35. Rogers, E. M. (1983). Diffusion of Innovation. New York: The Free Press.
36. Röling, N. (2009). Pathways for impact: scientists' different perspectives on agricultural innovation. International Journal of Agricultural Sustainablity, 7 (2), 83-94. http://dx.doi.org/10.3763/ijas.2009.0043

37. Schumpeter, J., Salin, E., & Preiswerk, S. (1980). Kapitalismus, Sozialismus und Demokratie. Munchen: Francke.
38. Smiths, R., Kuhlman, S., & Shapira, P. (2010). The Theory and Practise of Innovation Policy—an international research handbook (cited after EC-SCAR, 2012).
39. SUSTAINET EA. (2010). Technical Manual for Farmers and Field Extension Service Providers: Farmer Field School Approach. Nairoby: SUSTAINET East Africa. Retrieved from http://www.sustainetea.org.
40. Vaarst, M., Nissen, T. B., Østergaard, S., Klaas, I. C., Bennedsgaard, T. W., & Christensen, J. (2007). Danish Stable Schools for Experiential Common Learning in Groups of Organic Dairy Farmers. Journal Dairy Science, 90 , 2543-2554. http://dx.doi.org/10.3168/jds.2006-607
41. Wezel, A., Bellon, S., Dore, T., Francis, C., Vallod, D., & David, C. (2009). Agroecology as a science, a movement and a practice: A review. Agronomy for Sustainable Development, 29, 503-515. http://dx.doi.org/10.1051/agro/2009004

CHAPTER 8

Organic Farming and Sustainable Agriculture in Malaysia: Organic Farmers' Challenges Towards Adoption

Neda Tiraieyari, Azimi Hamzah, and Bahaman Abu Samah

8.1 INTRODUCTION

Environmentalists, ecologists, agricultural professionals, and policy makers are examining why the massive usage of chemicals in agriculture has led to soil and water pollution, loss of biodiversity, the destruction of natural habitats and many other negative consequences (Sadati, Fami, Asadi, & Sadati, 2010). Sustainable agriculture is at the heart of organic agriculture. Organic Farming (OF) is one of the agro-ecological approaches needed to grow enough food for the increasing population (Azadi et al., 2011). This approach minimizes external inputs such as chemical fertilizers, pesticides to produce non-toxic crops. Thus, it is less environmentally damaging and has much potential to produce more food, as a news release from the University of Michigan has explained. It is known to be an approach that aims to overcome some negative impacts of the Green Revolution on soil, water, landscape, and humans. According to (Partab, 2010) OF is an ecological agriculture that mostly depend on the management of ecosystems. This agricultural approach is not applicable just for developed countries but is also suitable for the developing world as well. In developing countries can contribute to socio-economic sustainability (Scherr & McNeely, 2008; Willer, Yussefi, & International Federation of Organic Agriculture Movements, 2004). For instance, it is claimed that OF can help rural development by contributing to tourism activities and generating employment and income to support local economies, especially in poorer countries (Hülsebusch, 2007; Scialabba, 2000).

While about 1 billion people in the developing countries are suffering from food, shortages and environmental degradation through unsustainable agriculture, yields in developing countries could increase by converting to OF (American Chemical Society, 2006). According to (Smolik, Dobbs, & Rickerl, 1995), OF is more profitable than conventional farming in the long-term. Studies conducted by (Badgley et al., 2007) have shown that OF can produce almost the same yield of conventional farming in developing countries. (Badgley et al., 2007) examined a global dataset of 293 examples, in order to compare yields of organic versus conventional methods of farming. Results showed that in developing countries, organic farmers produce 80% more than conventional farmers. (Badgley et al., 2007), also reported that OF could produce enough food and fiber to sustain the current human population without more land being cultivated. Another study by the (American Chemical Society, 2006) on 286 farm projects in 57 countries showed that organic farming not only protected the environment but also the yields increased by an average of 79 percent. According to Ann (2007), "OF can yield up to two or three times as much food as conventional farming on the same amount of land" which invalidates the assumption that OF cannot produce enough food and fiber to feed the world.

Scholars have argued much about the urgency of sustainable agriculture generally and OF in particular, with regard to agriculturally based countries such as Malaysia (Ahmad, 2001; Barrow, Weng, & Masron, 2009). In this country, almost 90% of Malaysian farmers in the food sector are small-scale producers for uneconomic sized farms, with a high cost of production, inputs and yield are low, and with low quality of products (Tiraieyari & Uli, 2011). The government's policy towards agriculture stresses increasing production to achieve food self-sufficiency. The Malaysian government has helped farmers with fertilizer subsidies to increase their production and improve their income (FAO, 2004). Therefore, the efforts for producing sufficient food and fiber for self-sufficiency have meant that Malaysia has adopted an intensive agricultural system and consequently has suffered environmental damage (Barrow et al., 2009). However, recently because of the growing awareness of the effects of unsustainable agriculture on their health and nature, the demand for organic products has risen among consumers considerably (Christopher, 2012). According to (Rezai, Mohamed, & Shamsudin, 2011) the rapid socio-economic development and the increasing standard of life has changed consumers' perceptions and awareness of organic products. As a result, the market for organic products in Malaysia has been developing quickly from few years back. Nonetheless, the supply of locally produced organic products is not enough to satisfy the increasing demand. Consequently, Malaysia needs to import organic production from other countries, especially from Australia, the U.S., and New Zealand (Christopher, 2012). In fact, the government has more appreciated the value of the agricultural sector and its contribution to the economy of country. Sustainable agriculture practices have also been lunched in the country in order to transform the agricultural sector in a sustainable manner. In the Third National Agriculture Policy (NAP3), organic agriculture was identified as a niche market opportunity for Malaysian small-scale producers (Ahmad, 2001). Even though the national agricultural policies are along with the standard of sustainable agriculture, but the current agricultural practices differ in terms of sustainability (Murad, Mustapha, & Siwar, 2008) and efforts made to promote OF in Malaysia have not generated acceptable results. A number of farmers have also been adopted sustainable agriculture in general and OF in particular practices are negligible. Malaysian farmers' challenges towards adoption of the programme may not be entirely clear for the policy makers. To our knowledge, few researchers have focused their study on unsustainable agriculture and the environmental damage caused by conventional farmers in Malaysia (Barrow, Chan, & Masron, 2010; Barrow, Clifton, Chan, & Tan, 2005; Barrow et al., 2009). In fact, little is known about organic farmers that adopted such practices in Malaysia. This study is designed to highlight the challenges of organic farmers in Malaysia.

8.1.1 ORGANIC FARMING IN MALAYSIA

In Malaysia, OF has a relatively young history (Christopher, 2012). It was begun by the Center for Environment, Technology and Development (CETDEM) in 1986 in a one ha in Sungai Buloh. In the mid-1990s, the country started to import organic products. The consumers for organic products were mainly cancer patients. In 1995 a number of commercial OF vegetable growers included more than 500 families (Wai, 1995). Malaysian government plans included encouraging small-scale producers to invest in OF as approach to increase their income, protect the environment and promote the country's exports. The NAP3 identified organic agriculture as a market opportunity, mainly for vegetable and fruits growers (Gunnar, 2007). In the eighth Malaysia Plan (2001 to 2005), the government aimed at increase of organic production by 250 ha (Wai, 1995). Government included the providing of assistance to farmers up to US$1,300 per ha in forms of infrastructure development. This also involved a certification scheme to cover the domestic market. Organic farming received government support through the establishment of national regulations for the Malaysian Organic Certification Program known as Sijil Organik Malaysia (SOM), which was launched in 2003 by the DOA to facilitate OF in Malaysia and to certify farms based on the requirements of the Malaysian standard MS1529:2001 (Kala et al., 2011). This standard, which is based on the Malaysian standard, sets the requirements for production to cover all stage of production. Moreover, the standard includes standards to control those hazards that affect the environment, food and workers' health and safety (Malaysia & Bahagian Pertanian, 2007). The scheme is open for participation by all farmers who are engaged in the primary production of fresh organic food products. The DOA is responsible for the implementation of the organic scheme. A group of trained agricultural officers has been assigned to carry out field inspection to verify that the farm operations or practices are in accordance with the organic standards. In 2002, the Minister of Agriculture noted that support services such as extensions, research, and development would be devoted for developing organic agriculture in Malaysia. In the ninth Malaysia Plan (2006-2010), the government targeted the OF, which was said to be worth more than US$ 200 million over 5 years. The Ministry of Agriculture planned to have 20,000 ha of (OF) by year 2010 and to increase local production by 4,000 ha per year.

In 2001, the DOA reported that there were only 27 organic farmers in the whole country with a total area of 131 ha. Currently, there are several privet organic farms in Malaysia. In 2010, it was about 42 certified holders occupying 1130 ha of land under OF system focusing on vegetable production, fruits, animal husbandry and aquaculture. In 2013 the DOA reported that there was a total of 89 farms occupying 1633.89 ha of land under OF and that 49 farmers have valid certification while 40 farmers had expired certification. The majority of organic farmers were in the Pahang states of Malaysia. OF is mostly restricted to vegetable growing and very few fruits are grown organically. Most organic fruits import from Australia, New Zealand, China, Korea and Japan. Local organic vegetables have a higher price; usually three times more than l conventional products due to the labor-intensive approach (Ahmad, 2001).

8.2 METHOD

This survey was designed to highlight the challenges that certified organic farmers are facing with regard to adoption of these practices. We used focus group interviews as a data collecting strategy. Qualitative data was collected from small groups of farmers that were successful in adopting organic technologies and in marketing their organic products. Data for this study was collected using focus group discussion (FGD) with organic farmers. Farm visits were also used to get a better vision of the problem. Focus group discussion (FGD) was conducted in March 2013.

8.2.1 LOCATION OF STUDY

The study was conducted in the district of Cameron Highland. The main reason why this area was studied is that it has been the most important productive region in producing vegetables in the country. It is a mountainous region of peninsular Malaysia with a total land area of 71,218 ha and mild temperatures and 5500 ha of the total land are devoted to agriculture. Due to the favorable climate, this area becomes the main producer of vegetable in Malaysia. Vegetable occupy (50%) of agricultural land, followed by tea plantation (40%), flower (7%) and fruits (2%) (Aminuddin, Ghulam, Abdullah, Zulkefli, & Salama, 2005). Currently there are 13 certified organic farmers in Cameron highland. We managed to get certified organic farmers through contact with director of organic farmers' association. The organic farmers' association is an independent small organization named the Cameron Organic Produce (COP) managed by representatives of the organization. The organization was established in 1996.

8.2.2 PARTICIPANT AND SAMPLING PROCEDURES

We conducted three focus group discussions with one group of organic farmers (n=6) from the Cameron Organic Produce (COP), one group of individual organic farmers (n=4) and directors and managers of COP as rich source of information (n=3). According to Stewart and Shamdasani (1990) there are no general rules for the best number of focus groups. They mentioned the rationale of working out the number of groups according to the homogeneity of the population, and the comfort of research application. Furthermore, they suggested that one focus group may well be enough. Regarding the size of focus group, Carey (1994) revealed that smaller groups were more manageable than bigger group. She states that the fewer people there are in the group, the greater the possibility that they will interact. In this study one of the researchers who involved in the project played the role of moderator. Millward (1995) suggests that the moderator should be directly involved in the project because they will be sensitive to the issues. The moderator explained the method and the purposed of research to the participants prior to commencing the interview. The use of tape recorder was also explained, and participants were informed that they can stop the interview/ discussion at any time they wish.

8.3 RESULTS

The following outline presents our findings on famers' challenges towards adopting (OF) in Cameron Highland.

8.3.1 LAND ISSUE

Farmers mentioned the land issue as their main challenge, especially for those who do not own the land, in order to start organic farming. The issue of Temporary Occupation Licenses (TOLs) was brought up in the Cameron highland in the early 1980s. Farmers are allowed to cultivate the land temporarily and the land is renewed annually by the government but the government as the owner can reclaim the land. Hence famers are not motivated to invest in land conservation when they never know how long they will be able stay on the land. Furthermore, farmers cannot get financial assistance or any kind of loan from the banks since the banks do not recognize TOL to give loans to the farmers. Farmers also reported delays in renewing their TOLs. The majority of small-scale producers in Cameron Highland are working on land which is held by the issue of Temporary Occupation Licenses (TOLs) by the government. Although the organic farmers we interviewed own the land and work on farms of less than 2 ha, these were mostly inherited from their parents. But due to land issue in Cameron Highland they cannot expand their farms and buy more land.

8.3.2 LABOUR SHORTAGES

Organic farmers in Cameron highland reported that labour shortages as the second most important challenge. Since their activities are done manually and organic technologies are mostly labour intensive, they have greater labour needs compare conventional farming methods. Due to the lack of local labour, organic farmers rely heavily on foreign labour to handle their work effectively. There is a great need for organic farmers in Cameron Highland to employ foreign workers, especially those from Bangladesh, Sri Lanka and Indonesia. In Malaysia, foreign workers in the agricultural sector must be directly employed by the government due to the immigration procedures and the need for working permits. The needs of foreign employees in the Malaysian agricultural sector cannot be denied. However, the recent government's policy has been to create more employment opportunities for locals and there has been a serious effort by the Malaysian government to reduce the country's dependence on foreign labor.

8.3.3 LACK OF TRAINING AND EXTENSION SERVICES

The farmers reported a lack of extension services for organic farming in Cameron Highland. There is no training in OF provided by the DOA or other government agencies to the farmers. Not much extension work has been done on transferring OF to the farmers. As a result organic farmers must seek information through multiple channels such as their colleagues, members of farmers' association, internet and attending workshop. The farmers reported that extension workers have little technical knowledge of organic agriculture. In other words, extension workers are not specifically trained in organic agriculture.

8.3.4 MARKETING

Farmers reported two main challenges in marketing their products in Malaysia. Firstly, the majority of people are not aware of the health benefits of consuming organic products, although demand for organic products has risen considerably among Malaysian people. Secondly, many people cannot afford to purchase organic products. Organic farmers have overcome these challenges partly by marketing their products through farmers' associations. The organic farmers' association establishes direct contact with special domestic buyers. The organic farmers' association in CH had also developed a specific website for selling their own products. Organic farmers also mentioned that exporting their products to a neighboring country such as Singapore is another challenge for them, although the demand for organic products in Singapore is very high. However, this country does not import organic products which are produced and certified to the SOM standard.

8.3.5 CERTIFICATION PROCESS

One of the main problems of the certification process is that it is extremely expensive. In addition, some unnecessary complexity increases confusion among farmers. Moreover, they reported the certification process takes too long for the government, at roughly 2-3 years. Organic farms have to be examined annually by a certified provider. Normally, members of the farmers' association build strong social relationship among themselves for developing and spreading information to overcome the complexity of the certification process.

8.3.6 GOVERNMENTAL SUPPORT

The farmers mentioned a lack of financial support from the government as their last but not their least challenge. In fact, organic farming requires significant on-farm and off-farm investment. Organic farmers need financial support to support their investments in soil conservation, production costs, labour costs, certification cost, and packing and storage facilities. According to the farmers, the government is not providing incentives for organic farmers. Mostly they receive help from NGOS.

8.4 DISCUSSION AND CONCLUSION

This article has reviewed the status of sustainable agriculture and organic farming in Malaysia. Sustainable practice has great potential in the country and therefore needs considerable support from the government. Maximizing production has been the most important goal for agricultural agencies and as a result governmental support for sustainable agriculture and organic farming has been limited. Organic farming and sustainable agriculture has not been practiced by the majority of farmers and the percentage of organic farmers in Malaysia is very small. Focusing on environmental awareness cannot secure conversion and it should not be the only approach offered by agricultural agencies to farmers. Government agencies should go beyond promoting such practices. The results of this study have also revealed challenges to organic farmers in Cameron Highland in terms of adoption of organic practices; include land issue, labour shortages, a lack of training and extension services and marketing, a lack of governmental support and the certification process. Among these the land issue appears to be the most serious barrier for farmers in converting to organic farming in Cameron Highland. Land tenure is critical to the adoption of such programmes. Obviously, tenant farmers would not be interested in investing in the land and go through a difficult conversion period without a strong guarantee of access to the land in later years. The unsolved land issue in Cameron Highland inhibits adoption of any sustainable agricultural practices. Organic agriculture is being promoted in Malaysia as a strategy to raise small-scale producers' income, protect the environment, and reduce food imports. However, from almost 2000 small-scale producers in Cameron Highlands who mainly work on farms of less than 2 ha only 13 adopted the program (Barrow, 2010). These 13 organic farmers own the land. One of the main ways Malaysian governments could influence adoption of the new method is through establishing property rights for small-scale producers. Wealthy farmers should not only be the target group for the program. Initially the DOA can begin with those who have stable forms of land tenure and can provide incentives and extension services to them. In addition, the Malaysian government should promote long-term rental contracts for small-scale producers and make provision to compensate farmers upon the termination of their contracts for investments made by them to improve the land. The Government should also target landowners and convince them of the importance of the land-conservation measures needed to obtain their support before promoting organic agriculture among small-scale producers.

It appears that small-scale producers rely on friends and media to receive information on organic farming. In order to promote the adoption of organic agriculture among small-scale producers, the government should support the adoption of the programme. The DOA should implement a policy of providing information on and research into organic farming. Agricultural extension has been essential in transforming the quality of information supplied to farmers. So the government should increase training for extension agents and specialists so that they can become recognized as useful sources of information on organic production. The DOA and other agricultural agencies should put organic farming on their priority programme list in order to facilitate the building of knowledge and human resources development. In other words, the success of this programme depends to a large extend on the training of farmers. The DOA should take steps to create adequate human resources and experts in research and extension should be one of its first priorities. An adequate number of well-trained extension agents on sustainable agriculture generally and OF specifically will play significant role in achieving sustainable agriculture in Malaysia. The DOA should facilitate the regular delivery of knowledge and experiences to farmers and provide them with direct incentives. Research, education, and extension efforts on sustainable agriculture and organic farming should be directed to organic agriculture in Malaysia.

The government can play a role in improving market access and helping organic farmers establish local marketing association. There are market opportunities for tropical countries such as Malaysia, especially for products that are not produced in Europe. Policy makers can also focus on the successful experiences of some Asian countries such as Japan, China, Korea and Thailand for Malaysia.

It is recommended that Malaysia needs to allocate resources and expertise to create organic research and development institutions/agencies. The government should consider providing direct incentives and some financial support to organic farmers. They should also facilitate certification process systems and hire foreign labour, in order for such projects to be successful.

REFERENCES

1. Ahmad, F. (2001). Sustainable agriculture system in Malaysia. In Regional Workshop on Integrated Plant Nutrition System (IPNS), Development in Rural Poverty Alleviation, United Nations Conference Complex, Bangkok, Thailand (pp. 18-20). Retrieved from http://banktani.tripod.com/faridah.pdf
2. American Chemical Society. (2006). Sustainable Farm Practices Improve Third World Food Production. Science Daily. Retrieved from http://www.sciencedaily.com/releases/2006/01/060123163315.htm
3. Aminuddin, B. Y., Ghulam, M. H., Abdullah, W. Y. W., Zulkefli, M., & Salama, R. B. (2005). Sustainability of Current Agricultural Practices Int. Cameron Highlands, Malaysia. Water, Air, & Soil Pollution: Focus, 5, 89-101. http://dx.doi.org/10.1007/s11267-005-7405-y
4. Ann, A. (2007). Organic farming can feed the world, U-M study shows. School of Natural Resources and Environment, University of Michigan. Retrieved June 26, 2013, from http://www.snre.umich.edu/newsroom/2007-07-10/organic-farming-can-feed-the-world-u-m-study-shows
5. Azadi, H., Schoonbeek, S., Mahmoudi, H., Derudder, B., De Maeyer, P., & Witlox, F. (2011). Organic agriculture and sustainable food production system: Main potentials. Agriculture, Ecosystems & Environment, 144(1), 92-94. http://dx.doi.org/10.1016/j.agee.2011.08.001

6. Badgley, C., Moghtader, J., Quintero, E., Zakem, E., Chappell, M. J., Avilés-Vázquez, K., & Perfecto, I. (2007). Organic agriculture and the global food supply. Renewable Agriculture and Food Systems, 22(2), 86. http://dx.doi.org/10.1017/S1742170507001640
7. Barrow, C. J., Chan, N. W., & Masron, T. B. (2010). Farming and Other Stakeholders in a Tropical Highland: Towards Less Environmentally Damaging and More Sustainable Practices. Journal of Sustainable Agriculture, 34, 365-388. http://dx.doi.org/10.1080/10440041003680205
8. Barrow, C. J., Clifton, J., Chan, N. W., & Tan, Y. L. (2005). Sustainable development in the Cameron highlands, Malaysia. Malay. Journal of Environmental Management, 6, 41-57.
9. Barrow, C. J., Weng, C. N., & Masron, T. (2009). Issues and challenges of sustainable agriculture in the Cameron Highlands. Malaysian Journal of Environmental Management, 10(2), 89-114. Retrieved from http://journalarticle.ukm.my/2290/
10. Carey, M. (1994). The group effect in focus groups: Planning, implementing, and interpreting focus group research. In J. Morse (Ed.), Critical Issues in Qualitative Research Methods (pp. 225-241). Sage Publications, London.
11. Christopher, T. B. S. (2012). Organic agriculture and food in Malaysia. Christopher Teh Boon Sung. Retrieved June 26, 2013, from http://christopherteh.com/blog/2012/02/organic-agriculture/
12. FAO. (2004). Fertilizer use by crop in Malaysia. Retrieved June 26, 2013, from http://www.fao.org/docrep/007/y5797e/y5797e00
13. Goh, C. S., Tan, K. T., Lee, K. T., & Bhatia, S. (2010). Bio-ethanol from lignocellulose: Status, perspectives and challenges in Malaysia. Bio resource Technology, 101(13), 4834-4841. http://dx.doi.org/10.1016/j.biortech.2009.08.080
14. Hanim, A. (2011). Agriculture becoming major contributor to GDP. The Star Online Business. Retrieved July 3, 2013, from file:///C:/Users/IPSAS-UPM/AppData/Roaming/Mozilla/Firefox/Profiles/dpu0nm1j.default/zotero/storage/ VGQ7FJ3T/Agriculture%20becoming%20major%20contributor%20to%20GDP.htm
15. Hülsebusch, C. (2007). Organic Agriculture in the Tropics and Subtropics-Current Status and Perspectives. kassel university press GmbH.
16. Kala, D. R., Rosenani, A. B., Fauziah, C. I., Ahmad, S. H., Radziah, O., & Rosazlin, A. (2011). Commercial Organic Fertilizers and their Labeling in Malaysia. Management Journal of Soil Science, 15, 147-157. Retrieved from http://www.msss.com.my/mjss/Full%20Text/Vol%2015/kala.pdf
17. Kamaruddin, R., & Masron, T. A. (2010). Sources of growth in the manufacturing sector in Malaysia: Evidence from ardl and structural decomposition analysis. Retrieved from http://www.academia.edu/download/30908574/AAMJ_15.1.6.pdf
18. Malaysia and Bahagian Pertanian. (2007). Standard Skim Organik Malaysia (SOM) Malaysian organic scheme. Putrajaya: Jabatan Pertanian Malaysia.
19. Millward, L. (1995). Focus groups. In G. Breakwell, S. Hammond, & C. Fife-Schaw (Eds.), Research Methods in Psychology (pp. 274-292). Sage Publications, London.
20. Mohamed, M. S. (2007). Status and Perspectives on Good Agricultural Practices in Malaysia. Food and fertilizer technology center. Retrieved from http://www.agnet.org/library.php?func=view&id=20110725102453
21. Murad, M. W., Mustapha, N. H. N., & Siwar, C. (2008). Review of Malaysian Agricultural Policies with Regards to Sustainability. American Journal of Environmental Sciences, 4(6), 608-614. http://dx.doi.org/10.3844/ajessp.2008.608.614

22. Partab, T. (2010). Merging organic farming sector in Asia: A synthesis of challenges and opportunities. In T. Partab, & M. Saeed (Eds.), Organic Agriculture and Agribusiness: Innovation and Fundamentals. The Asian Productivity Organization. Retrieved from http://www.apo-tokyo.org/publications/files/agr-22-oaa.pdf
23. Rezai, G., Mohamed, Z., & Shamsudin, M. N. (2011). Malaysian consumer's perceptive towards purchasing organically produce vegetable. In 2nd International Conference on Business and Economic Research, Holiday Villa Beach Resort and Spa, Langkawi, Kedah, Malaysia. Retrieved from http://www.internationalconference.com.my/proceeding/2ndicber2011_proceeding/310-2nd%20ICBER%202011%20PG%201774-1783%20Malaysian%20Consumers%20Perceptive.pdf
24. Sadati, A., Fami, H. S., Asadi, A., & Sadati, A. (2010). Farmer's Attitude on Sustainable Agriculture and its Determinants: A Case Study in Behbahan County of Iran. Research Journal of Applied Sciences, Engineering and Technology, 2(5), 422-427.
25. Scherr, S. J., & McNeely, J. A. (2008). Biodiversity conservation and agricultural sustainability: Towards a new paradigm of "eco agriculture" landscapes. Philosophical Transactions of the Royal Society Biological Sciences, 363(1491), 477-494. http://dx.doi.org/10.1098/rstb.2007.2165
26. Scialabba, N. (2000). Factors influencing organic agriculture policies with a focus on developing countries. In IFOAM 2000 Scientific Conference, Basel, Switzerland (pp. 28-31). Retrieved from http://www.fao.org/docs/eims/upload/230159/BaselSum-final.pdf
27. Smolik, J. D., Dobbs, T. L., & Rickerl, D. H. (1995). The relative sustainability of alternative, conventional, and reduced-till farming systems. American Journal of Alternative Agriculture, 10(1), 25-35. http://dx.doi.org/10.1017/S0889189300006081
28. Stewart, D., & Shamdasani, P. (1990). Focus groups Theory and Practice. Sage Publications, Beverly Hills, CA.
29. Tiraieyari, N., & Uli, J. (2011). Sustainable Agriculture in Malaysia: Implication for Extension Workers. Journal of American Science, 7(8). Retrieved from http://www.jofamericanscience.org/journals/am-sci/am0708/018_6034am0708_179_182.pdf
30. Wai, O. K. (1995). The role of agriculture and rural development in Malaysia. IV. National study: Malaysia IV. Retrieved July 4, 2013, from http://www.unescap.org/rural/doc/oa/Malaysia.PDF
31. Willer, H., Yussefi, M., & International Federation of Organic Agriculture Movements. (2004). The world of organic agriculture: Statistics and emerging trends. Bonn, Germany: IFOAM.

CHAPTER 9

Are Organic Standards Sufficient to Ensure Sustainable Agriculture? Lessons From New Zealand's ARGOS and Sustainability Dashboard Projects

Charles Merfield, Henrik Moller, Jon Manhire, Chris Rosin, Solis Norton, Peter Carey, Lesley Hunt, John Reid, John Fairweather, Jayson Benge, Isabelle Le Quellec, Hugh Campbell, David Lucock, Caroline Saunders, Catriona Macleod, Andrew Barber, and Alaric Mccarthy

9.1 INTRODUCTION

Maintaining biodiversity and other ecosystem services to sustain efficient food and fibre production is one of the greatest challenges facing humanity (Millennium Ecosystem Assessment, 2005). Efficient industrialised agriculture,

---

powered by energy and nutrient subsidies and technology, helps secure human wellbeing by providing "provisioning services" (efficient and sustainable production of food and fibre). However it has also weakened nature's ability to deliver other key regulating and supporting ecosystem services, e.g. purification of air and water, protection from disasters, and nutrient cycling. "Cultural ecosystem services" underpin connection to place, community support, land stewardship values, local economies, transfer of knowledge, and the identity of farmers. These cultural services provide the incentives and enhance capacity to sustain and adapt coupled social and ecological systems. All types of ecosystem services are required to capture new opportunities and counteract challenges such as climate change, peak oil, globalisation of markets, biosecurity risks and transgenic organisms (Darnhofer et al., 2011; Pretty et al., 2010).

Market assurance and certification schemes have emerged as a global response to encourage and reward sustainable agriculture and inform consumers (Campbell et al., 2012b). Such schemes often stipulate best farming practices and many establish explicit standards, sustainability assessments, monitoring and audits that seek to future-proof ecosystem services in production landscapes. They are designed to assure consumers and regulators that the food and fibre has been produced in an ethical and sustainable way, and that foods are safe and nutritious to eat. "Organic Agriculture" is one of the very earliest and well recognised of such market accreditation schemes. There are now scores of other frameworks, standards and certification schemes that purport to enhance the economic resilience and sustainability of production. Some adopt elements of Integrated Pest Management, or more broadly 'Integrated Management', that seek to reduce and optimise the chemical applications and farm inputs in general and include whole farm management systems that promote efficient use of resources and land. They increasingly incorporate social and governance dimensions of ethical farming (e.g. good labour relations, animal welfare, and broader biodiversity care). For instance, the United Nations Food and Agriculture Organisation has recently promulgated the Sustainability Assessment of Food & Agriculture (SAFA) in an attempt to harmonise this growing and diverse range of sustainability assessment schemes (FAO, 2013). This raises an important question that we examine in this paper: Are organic standards sufficient to secure ecosystem services in the broader way that SAFA and other frameworks are now promulgating as necessary to ensure sustainability and resilience of farming?

This paper begins by briefly reviewing some of the results of the ARGOS project, a nine-year longitudinal study of organic and other farms in New

Zealand. Next we present a broad 'gap analysis' between the Organic standards and principles and the new dimensions of sustainability incorporated into the SAFA. Then we describe the New Zealand Sustainability Dashboard project as an example of a tool that could close the gaps between organic standards and IM frameworks like SAFA. We conclude by examining options for the organic movement to better protect and enhance ecosystem services and secure its premiere market position for delivering sustainable and ethical food and fibre production.

9.2 DOES ORGANIC FARMING DELIVER MORE, FEWER OR DIFFERENT ECOSYSTEM SERVICES?

The Agriculture Research Group on Sustainability (ARGOS) was a transdisciplinary project measuring the ecological, economic and social outcomes from over 96 farms in New Zealand between 2004 and 2012. The project sought well-replicated and long-term research of whole working farms from different land use intensities. It compared economic, social, environmental and farming practice outcomes between Organic, "integrated management" (IM) and "conventional" orchards and farms (Campbell et al., 2012a, b). The Organic panels were certified as following organic standards. The IM panels had adopted a market assurance scheme that incorporated several principles of best farming practice, including elements of integrated pest management and optimisation of farm inputs. The "conventional" farmers did not adhere to any collective market assurance protocols. Examination of several hundred parameters tested an overarching null hypothesis of the study: Ho: economic, social and environmental outcomes are the same for organic, integrated management and conventional farming systems.

One commercial farm or orchard from each available farming system was chosen in each of 12 clusters for each sector ('kiwifruit, 'sheep/beef' and 'dairy') spread throughout New Zealand (Campbell et al., 2012a). Clustering ensured that soils, topography, climate, ecological constraints and rural community drivers were similar for each farming approach in a given vicinity. Spreading the clusters ensured a more representative test of the null hypothesis across several regions of New Zealand. There are no IM dairy farms and all conventional kiwifruit orchards have converted to IM in New Zealand, so only the sheep/beef sector had all three farming systems available for comparison. General Linear Modelling used a blocked design to remove the effects of cluster from statistical tests of the main effects of farming system on outcomes. The dairy farm panels were monitored before conversion of half of them to organic farming, so in that case we could use a Before-After-Control-Impact approach to test whether adoption of certified organic standards causes changes in outcomes. Sheep/beef and kiwifruit farms had converted to organic or IM farming systems long before the ARGOS study began, so any observed differences in current performance of the farms will only reflect their farming system practices if we can safely assume that sustainability indicators and performance were about the same before their conversion to organic or IM methods occurred. ARGOS therefore provided a well-replicated and relatively long-term comparison of outcomes on real working farms following different market assurance protocols with outcomes on a reference group of non-assured ("conventional") farms.

9.2.1 PROVISIONING SERVICES ARE REDUCED ON ORGANIC FARMS

Farmers primarily tune production landscapes for efficient production of food and fibre: the key provisioning service. A consistent finding of the ARGOS project was that production per hectare of land was much reduced in organic farming. For example, dairy farms converting to organic showed a consistent decline in production (milk solids/ha) relative to conventional dairy farms over a five-year period (Figure 9.1). The largest difference in production was observed once converting farms became certified as organic growers, with organic farms producing only 69% of that of their conventional counterparts. Milk production was already lower on converting farms before they sought organic certification. This suggests that there was something about those farming families, their land or their existing farm practices before they actually formally adopted certified organic methods that resulted in lower production. This serves a clear warning that many organic vs non-organic farm outcome comparisons may provide only quasi-experimental evidence that changes in ecosystem services including yields are caused by the organic farming practices themselves. A formal experiment would require random allocation of families and land to each panel, whereas in real life the existing orientation of the farmers or even characteristics of their land or economy may have predisposed some to go organic or IM, and others to remain conventional. Our results demonstrated that a mix of both predetermined and causally driven organic farming practice effects caused lower production because initial differences from conventional colleagues greatly increased as dairy practices consolidated and certification was conferred.

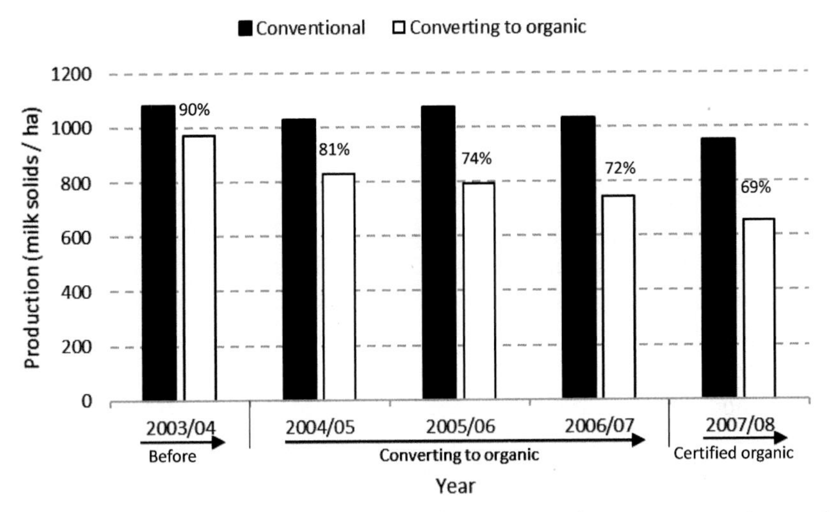

FIGURE 9.1 Annual production of dairy farms (milk solids / ha) for conventional farms and farms converting to organic, from the ARGOS project. Percentages of organic (converting to) production relative to conventional production is indicated for each year (Campbell et al., 2012a). Production measures have not been adjusted for land used to produce feed supplements imported from other farms.

Comparative provisioning efficiency between sectors can best be summarised by comparing the gross energy outputs and inputs per hectare of production land. A 24.5% reduction in energy production per hectare was observed on ARGOS organic green kiwifruit orchards compared to IM counterparts, and organic sheep and beef farmers produced on average 17.5% and 29.1% less energy than their IM and conventional counterparts respectively (Norton et al., 2010). The same general pattern for reduced production per hectare has been observed in organic systems across the board when compared with more intensive agricultural systems that drive increased production by imports of ecological and energy subsidies (e.g. Sato et al., 2005; Rozzi et al., 2007). However, energy inputs (from fertilisers, pesticides, supplementary stock feed and electricity) to organic ARGOS farms were also much reduced compared to IM and conventional farms, so they rely less on ecological and energy subsidies for production. The net efficiencies of production from an energy point of view ("Energy Return on Investment", EROI) were therefore remarkably similar between all farming systems within the sheep/beef sector; but 13.4% less efficient on organic compared to conventional dairy farms; and 12.5% less efficient on organic compared to IM green kiwifruit. From an overall energy systems efficiency point of view then, organic ARGOS farms were of similar or slightly reduced in efficiency. However, if efficiency is calculated purely as production per hectare of land used, organic farming would be judged as a far less efficient way of delivering provisioning services.

Two broad 'land allocation' paradigms have been promulgated for provisioning a growing world human population without undermining ecosystem services: a "land-sparing" approach promotes more intensive farming of land that is tuned for maximum productivity so that more land can be protected (often reserved) for other services such as biodiversity conservation (Lindenmayer et al., 2012); a "land-sharing" approach promotes farming practices that maintain natural capital and all the ecosystem services from the same land that produces food and fibre. Along this continuum, organic agriculture is potentially less effective in a land-sparing strategy because reduced productivity on farmland may trigger more conversion of ecosystem service refuge areas to farmland or forestry. However organic farming will enhance land-sharing outcomes if it enhances regulating, supporting and cultural services in the production spaces (fields) of farming landscapes. This underscores that judgements about net benefits of organic production compared to non-organic approaches are scale dependent and coupled to an underlying land allocation model for maintenance of ecosystem services. More research is needed to test whether lower production on organic farms indeed reduces land-sparing, or whether any such environmental deficit is more than made up for in biodiversity benefits from land-sharing.

9.2.2 ENHANCING BIODIVERSITY FOR SUPPORTING AND REGULATING CULTURAL SERVICES

Organic farming standards traditionally concentrated on: prohibiting the use of xenobiocides and xenobiotic chemicals as inputs into food; only allowing naturally occurring (eobiotic) fertilisers (synthetic nitrogen fertilisers are thus prohibited) and other "inputs"; banning transgenic and similar technologies and their products; increasingly restricting nanotechnology, within a framework that is focused on enhancing soil health and maintaining the 'wholeness' of food thus produced. There is now a substantial body of research showing that this can affect the abundance of pests, weeds and beneficial biodiversity in direct ways.

ARGOS found a greater variety of plants growing under shelterbelts (Moller et al., 2007) and higher species richness and abundance of invertebrates within the production areas (eg. Todd et al., 2011) of Organic compared to IM Kiwifruit orchards. Higher numbers of predators, parasitoids, herbivores, fungivores and omnivores in the organic orchards compared with those under IM are expected to result in more resilient ecosystem services in the organic orchards. The emergence of indirect effects in ecological food webs is of particular interest: might enhanced biodiversity or other ecosystem changes sufficiently substitute for the regulation services normally provided by chemical applications on conventional and IM farms? If so, more biologically efficient, inexpensive, practical and safe production can be expected from organic farming.

Biodiversity makes ecological systems adaptable and resilient to biophysical changes in production landscapes by supporting and regulating ecological processes needed for production of food and fibre. Community ecology has repeatedly emphasised that some species ('keystone species') have inordinate effects on other species in food webs, and some ('ecosystem engineers') are pivotal in creating habitat for whole new foodwebs and ecological processes. For example, earthworms comprise a major component of the animal biomass (non microbial) in soils and contribute to a range of ecosystem services through pedogenesis, development of soil structure, water regulation, nutrient cycling, assisting primary production, climate regulation, pollution remediation and cultural services (Kopke, 2015, this volume). The ARGOS study revealed higher earthworm density in organic kiwifruit orchards, but there was no evidence of them differing between farming systems for the dairy or sheep & beef sectors (Figure 9.2).

The ARGOS project assessed whether orchards managed under an organic system supported higher bird density and diversity than those under two different IM systems (Gold and Green Kiwifruit). Birds were researched because they are often 'top predators' in food chains (and thereby sensitive to ecosystem change), relatively easy to monitor, conspicuous and loved by many consumers. This makes them potential "Market Flagship" species for promoting sustainable farming practices and ethical purchasing by consumers (Meadows, 2012). Higher densities of all New Zealand native bird species (insectivores and nectar-feeders) were detected on orchards managed under organic systems, relative to IM orchards (MacLeod et al., 2012). This lends support to the hypothesis that organic farming systems sustain enhanced biodiversity compared to non-organic counterparts (Bengtsson, Ahnstrom & Weibell, 2005; Hole et al., 2005). However the introduced bird species were an order of magnitude more abundant on the orchards than native species and there was no evidence that their abundance differed between farming systems. The New Zealand public have an overwhelming preference for conserving native and endemic species rather than introduced ones, mainly because the native biota are closely embraced as part of New Zealand's national identity and conservation responsibilities. However European consumers of New Zealand produce may be most concerned by the support of their own threatened farmland species that have been introduced to New Zealand and flourish there. This demonstrates a need for a more nuanced focus on particular biodiversity that might have particular functional roles or particular biocultural significance in agriecosystems rather than a simple binary expectation that organic agriculture enhances biodiversity across the board.

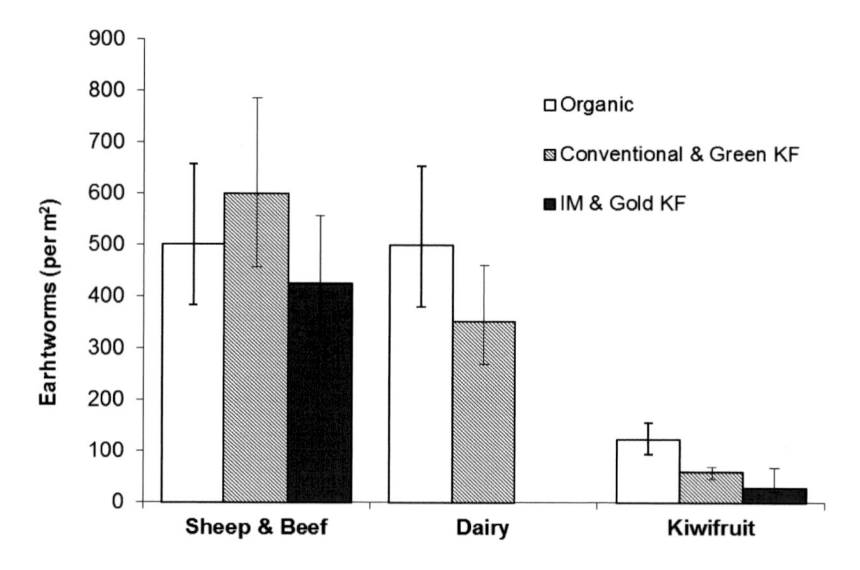

FIGURE 9.2 Earthworm density under different management systems for 36 sheep/beef, 24 dairy and 36 kiwifruit farms. Note that 'integrated management' was not available for the dairy farming sector and both Green and Gold Kiwifruit were grown under IM protocols. The error bars depict ± 1 standard error (Sources: after Carey et al., 2010).

7.2.3 A NEED TO MANAGE MORE THAN FARMING INPUTS

The above examples from the ARGOS project lead us to emphatically endorse the calls by Barberi (2015, this volume) and Niggli (2015, this volume) to focus on functional biodiversity. But we go much further to stress that further enhancement of sustainability of organic agriculture depends on finding and then managing the drivers of variation in those important animals and plants and key social-ecological systems processes. For example, pesticide loadings and woody vegetation cover proved to be more influential predictors of native bird densities than 'management systems' on New Zealand kiwifruit orchards: native bird density was lower where more pesticides were applied and higher on orchards with more woody vegetation (MacLeod et al., 2012). The presence of woody vegetation, while not considered in organic standards, provide vital ecological refuges and habitat for native New Zealand biota (Moller et al., 2008). We expect a synergistic interaction where the benefits of low toxicity of farm inputs will lift the average native bird abundance all the more above that of its non-organic counterparts if diverse and extensive woody vegetation is also retained. If organic farming actively promoted or even required provision of more woody vegetation on farms and orchards, we predict even higher density of birds would be found on organic farms. Another ARGOS example concerned spiders and beetles that provide important ecosystem services on dairy farms. Organic dairy farms and fenced shelterbelts supported 40% and 67% higher densities of spiders than conventional dairy farms and unfenced shelterbelts, respectively (Fukuda et al., 2011). Shelterbelts of native plant species supported higher species richness of native spiders and beetles than shelterbelts of exotic plants. So conversion to organics lifts biodiversity to some degree, but a combination of organic methods, fencing off shelterbelts and planting more native tree species in shelterbelts will provide all the more ecosystem services and biodiversity conservation on New Zealand dairy farms.

7.2.4 SUSTAINABLE INTENSIFICATION: MIGHT ORGANIC FARMING BE PARTICULARLY BENEFICIAL IN MORE INTENSIVE AGRICULTURE?

There is a clear need to transcend research from simple tests for significant differences in outcomes from organic, IM and conventional farming to testing larger scale hypotheses about the size, direction and reason for differences in ecosystem services between farming systems. An example is to test whether aspects of organic farming ameliorate the unwanted effects of landuse intensification. The ARGOS team proposed a second meta-hypothesis H1: Differences in economic, social and environmental outcomes between organic, integrated management and conventional farming are greater for more intensive farming sectors and farms. This emerged from ecological first principles – the higher the rate of application of ecological subsidies (i.e. anthropogenic subsidies of materials from outside an ecosystem's boundary, Pilati et al., 2009) such as artificial fertilisers and suplementaty feed for livestock, the greater the alteration of local ecology through immediate and direct ecological disturbance effects. Organic restrictions might lessen the force of such subsidies partly by their more benign nature and partly indirectly because organic farms are generally less intensive operations (reduced stocking rates, less extraction of nutrients and materials, lower productivity as seen in Figure 1 and EROI comparisons).

We had insufficient replication of sectors to fully test this intensification hypothesis, but preliminary observations are consistent with it. For example, the relative effect of farming system on earthworm abundance was much greater in the most intensive sector (Kiwifruit) than the next most intensive farming (dairy), and there was no evidence of a difference between systems in the least intensive sector (sheep & beef). Similar interactions between sector intensity and soil structure and its macronutrients were observed (Carey et al., 2010). Rudimentary binary comparisons of organic and non-organic outcomes abound in the literature, but so far they have not led on to testing higher order drivers why these differences occur, or why they are larger in some agriecosystems than in others. Halberg (2015, this volume), Vaarst (2015, this volume) and Heckman (2015 this volume) have all empahsised the need for 'eco-functional intensification'. If the ARGOS intensification meta-hypothesis is true, organics has special value in supporting ongoing intensification of agriculture without damaging ecosystem services.

9.2.5 ORGANIC AGRICULTURE AS AN AGENT FOR CHANGE: A ROLE FOR CULTURAL SERVICES

Our analysis thus far has mainly concerned ecological dimensions of ecosystem services. However discovery of the social and economic drivers of farming practice are also fundamentally important for sustaining coupled social-ecological systems (e.g, Rosin et al., 2008, Campbell et al., 2012b). The long term resilience and sustainability of agriculture depends on learning and adapatability (Vogl 2015, this volume). Transformation of agriculture to protect and enhance ecosystem services will depend on direction, motivation, "opportunity to perform" and ability (Tuuli, 2012; CEO Group, 2015). This means that farmers and policy makers will need an awareness of the need to change, the values and motivation to act in beneficial directions, and the capacity to make the required changes. Cultural ecosystem services include the nonmaterial benefits people obtain from ecosystems through spiritual enrichment, cognitive development, reflection, recreation, and aesthetic experiences (Millenium Ecosystem Assessment, 2005). We consider cultural services as potentially crucial for underpinning adaptability by building a sense of place and responsibility to other places and people, knowledge of the need and options for change, and forming core values to motivate change or strike balances and trade-offs between short term economic rewards and land care.

The ARGOS researchers used both formal Qualitative Analysis methods of semi-structured interviews and nationwide questionaniares (Fairweather et al., 2009a) to explore individual farmers' economic, social and environmental orientations (Table 9.1). Organic farmers displayed a much broader social and environmental 'breadth of view', were more likely to innovate, and were less focussed on economic success than their non-organic counterparts (Table 9.1). All these differences will make organic growers more aware of threats and opportunities for sustainability, and perhaps more ready to change when needed.

TABLE 9.1 Relative orientations of organic and non-organic farmers to four aspects of farming. A score between +1 (strong support) to – 1 (strong aversion) was determined by a Factor Analysis of each farmers' answers to a nationwide survey. (Sourced from Hunt et al., 2011).

Orientation	Non-organic (n= 338)	Organic (n = 157)	t-Test significance
Economic Focus (relative importance of economic success of their farm)	+0.07	-0.15	0.034
Social Breadth of View (relative contribution of their farming to wider society benefits)	-0.17	+0.37	<0.001
Environmental Breadth of View (relative importance to consider effects of their farming beyond their own land)	-0.16	+0.35	<0.001
Innovation Likelihood (relative willingness to experiment with their farming practice)	-0.21	+0.45	<0.001

9.3 WHAT SHOULD ORGANIC FARMING BE COMPARED AGAINST?

Much of the literature on organic agriculture presents binary comparisons of organic farming outcomes and their provision of ecosystem services compared to non-organic farming. The ARGOS results that we briefly summarised above emphasise the danger in such simplified binary comparisons: a rapidly growing group of IM farmers are adopting market accreditation and monitoring schemes to fine-tune their farming practice in ways that purport to be more sustainable. Outcomes from these IM growers are sometimes quite different from so-called "conventional' farmers. For example, in the ARGOS project, macoinvertebrate communities and ecosystem functioning were negatively impacted in streams running through conventional sheep/beef farms, but there was no evidence of them being different in Organic and IM farms (Magbanua et al., 2010).

The results of qualitative analysis of interviews (Campbell et al., 2012b) and the responses of IM, Organic and Conventional growers in a nationwide survey to questions about environmental, social and economic dimensions of farming sustainability (Figure 9.3) both emphasised that IM growers are different from conventional ones. The IM growers were not just intermediate between organic and conventional (had they been, the multidimensional scaling diagram would have approximated E in Figure 9.3). Instead they viewed farming in very different ways from both conventional and organic growers. Differences between the panels were relatively less for financial and social orientations (the same conclusion is demonstrated in Table 9.1), but organic farmers were particularly distinct in orientation to environmental and production concerns. We do not know what drives the differences already evident nor, potentially more crucially, how they might change in future because of the engagement of IM farmers in the market accreditation and sustainability best practice monitoring frameworks. Clearly there are many clusters of "greeness" in orientation within all types of farming approaches and the way these are influenced by market accreditation and reward is an important dynamic for guiding the way the organic movement positions itself in markets and as environmental friendly farming advocacy (Fairweather et al., 2009b; Campbell et al., 2012b). In the meantime we urge researchers and market advocates of organic agriculture to not simply lump all non-organic farmers into one pool, especially since the eco-verification and wider sustainability claims of the IM farmers could undermine the premiere and historical monopoly of market assertions that organic agriculture certification gaurantees sustainable and ethical production.

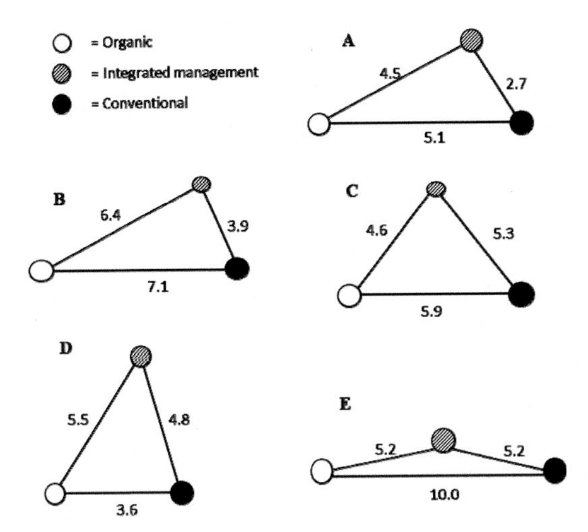

FIGURE 9.3 Multidimensional scaling to measure differences in the way New Zealand organic, IM and conventional farmers answered questions about different dimensions of farming. The numbers on each diagram are 'multivariate distances', a measure of how distinct the farmers from each farming system were in the responses to the same questions A: Production performance (9 questions), B: Environmental performance (17 questions), C: Social indicators (14 questions), D: Financial indicators (11 questions). E: a hypothetical example of IM as an intermediary between organic and conventional farming systems. (Source: The questionaire results are described by Fairweather et al. 2009a, and the multidimensional scaling is an unpublished analysis by Lesley Hunt.)

9.4 ON WHAT BASIS SHOULD STAKEHOLDERS COMPARE SUSTAINABILITY OF FARMING SYSTEMS?

In view of the rapidly rising prominance of the IM and market assurance farming protocols that are making sustainability claims, we sought to measure the degree of congruence and divergence between their tenets for ensuring sustainability and those incorporated into organic farming. The organic 'brand' is now synonymous with the organic standards, i.e., the 'rules' of organic farming systems. Traditionally organic sustainability claims are therefore based primarily around assumption that restricting the nature of farm inputs will protect and enhance ecosystem services and produce safer and higher quality food and fibre within a more ethical production system. More recently the standards have been mapped to and endorse four core 'IFOAM principles' (IFOAM, 2005): Health—Organic agriculture should sustain and enhance the health of soil, plant, animal, human and planet as one and indivisible; Ecology—Organic agriculture should be based on living ecological systems and cycles, work with them, emulate them and help sustain them; Fairness—Organic agriculture should build on relationships that ensure fairness with regard to the common environment and life opportunities, and; Care—Organic agriculture should be managed in a precautionary and responsible manner to protect the health and well-being of current and future generations and the environment. A comprehensive summary of the standards and principles (which we will henceforth collectively refer to as 'organic norms') is found in the Common Objectives and Requirements of Organic Standards (COROS). We chose to compare organic standards and principles with the FAO's SAFA framework because the latter is a recent, comprehensive and broadly applicable set of sustainability principles that attempts to integrate the features of a large number of IM and market assurance approaches.

9.4.1 HOW MANY OF SAFA'S SUSTAINABILITY CRITERIA ARE COVERED BY ORGANIC PRINCIPLES?

We searched for a match between each SAFA indicator and its description with the IFOAM 2014 and BioGro New Zealand organic standards. A five-point mark ranging from 0% (no correspondence), 25%, 50%, 75%, to 100% (complete correspondence) was scored for each SAFA indicator. Scoring was conducted by the lead author who has 24 years' experience of working with organic standards internationally to help make it as consistent and accurate as possible.

Figure 9.4 replicates the radar charts that are commonly used by SAFA to depict performance at each spoke of a "wagon wheel" that depicts a family of criteria required for sustainability. The inner red zone represents failure of compliance when used in real SAFA assessments, but in our case we use it to show 0% congruence of the organic standards with SAFA requirements. We equate the inner and outer margins of the next amber zone with 25% and 49% congruence, and so on outwards until 100% congruence is plotted against the outer margin of the deep green zone around the wheel's perimeter. The dark line in Figure 4 represents average congruence of organic standards for several indicators within each SAFA sustainability dimension.

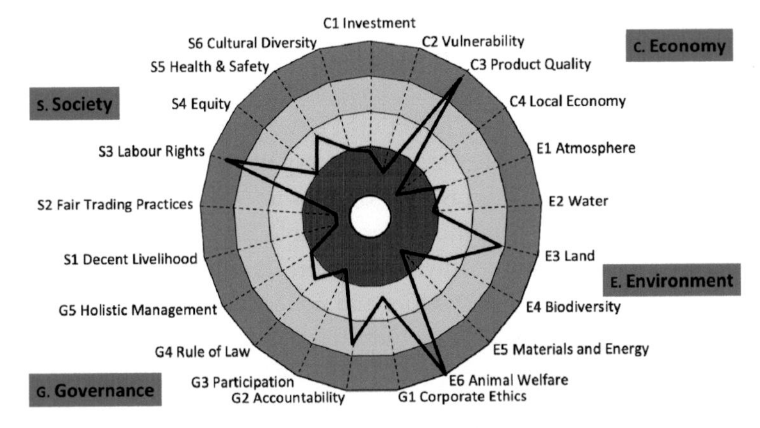

FIGURE 9.4 Sustainability scores of the organic standards when using FAO's SAFA criteria for sustainable food and fibre production. The black line indicates the degree of congruence between organic and SAFA sustainability criteria (the further the black line is from the centre the greater the congruence). Successive zones indicate grades of sustainability performance from deep green (most sustainable) around the outside to red (least sustainable) in the inner core. The SAFA framework has been customised to better meet New Zealand conditions by the Sustainability Dashboard.

The organic standards are almost completely in agreement with SAFA on issues of Product quality & information, Animal welfare and Labour rights; but organic standards are virtually silent on the need for Fair Trade, providing a decent livelihood, contributing to local economy and minimising reliance on materials and energy. Even aspects of environmental and land care (like biodiversity, water and atmosphere) that are explicitly required in SAFA assessments are only partially embraced by the organic standards.

Our overall average score for congruence (the average distance of the dark line from the centre of Figure 9.4) was 36%. It is important to remember that just because the organic standards do not fully cover a given sustainability criterion (and so score 0% or 25% if partially covered), many of the organic farms may nevertheless be performing very well on that dimension of sustainability (indeed our ARGOS examples in Figures 9.1-9.3 and Table 9.1 above suggest this is the case). Our aggregated score would only measure performance if the organic farm was fully achieving the explicit requirements of organic standards and no more. The comparatively low overall score simply emphasises that SAFA and many similar sustainability assessments are including a much wider set of necessary and quite explicit conditions than those required for meeting organic standards and principles.

There is a remaining emphasis on organic input restrictions: 47% of 90 COROS standards are framed in terms of farming input restrictions, 35% concern more general principles and outcomes, and 18% regulate internal consistency of the organic standards. The IFOAM principles are cast in such general and abstract terms that they are difficult to interpret and judge in terms of day-to-day farming decisions, whereas rules on organic farm input restrictions are precise, measureable and voluminous. For example, BioGro NZ, one of the two New Zealand export organic certifiers, covers the six COROS items on fairness, respect and justice, equal opportunities and non-discrimination in just a third of a page (132 words) of its 2011 certification standards, yet the "Directory of BioGro Certified inputs for producers" 2011 for facility management, dairy, crops, bee keeping, livestock soil and seeds, lists 251 different types of inputs, with some inputs having multiple individual approved suppliers, e.g., fish fertilisers list 88 different fish fertiliser products, with the directories covering 26 pages. AsureQuality, the other organic certifier in New Zealand, did not include a section on social justice until its 2013 (No 5) version of its standards and devotes just 317 (0.7%) words out of 43,782 to social justice.

9.4.2 HOW MANY OF THE ORGANIC STANDARDS ARE COVERED BY THE SAFA SUSTAINABILITY CRITERIA?

We then used the same scoring methods to perform the reverse comparison: how well would a farmer that is fully meeting SAFA performance criteria score if judged against organic standards? Standards do not have the equivalent of SAFA indicators. Instead they are more akin to a legal document with a large number of specific details. Therefore the COROS were used to undertake the comparison. COROS, also called the "The IFOAM Standards Requirements", is designed for use in international equivalence assessments of organic standards and technical regulations and provides the basis for assessing equivalence of standards for inclusion in the IFOAM Family of Standards.

For each 'Objectives and Requirements' in COROS we estimated that a fully compliant SAFA farmer would on average meet 74% of the organic norm requirements. An excellent non-organic farmer, according to SAFA criteria, performs well in terms of the requirements for organic farmers to be systems oriented, minimise pollution and land degradation, protect animal welfare and health, and act with fairness and respect (Figure 9.5). However, more stringent requirements on organic farmers for long-term and biologically-based soil management, avoiding synthetic inputs, and especially in avoiding unproven and unnatural technologies remain as points of difference in organic farming (Figure 9.5). These points of difference are reflected in very specific requirements for organic growers to avoid transgenic organisms, irradiation, certain breeding technologies and nanotechnology (Figure 9.6).

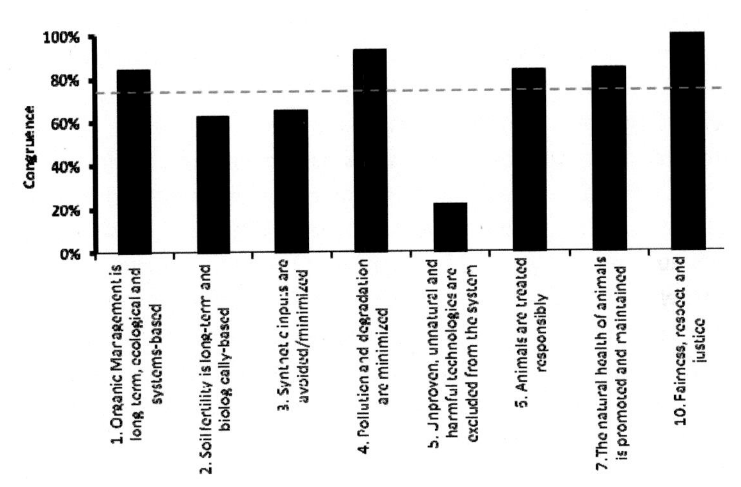

FIGURE 9.5 Congruence scores in higher order themes of a producer that is fully compliant with FAO's SAFA when judged against the IFOAM organic standards. The dashed line indicates the average degree of congruence (74%) for 78 specific requirements of organic production.

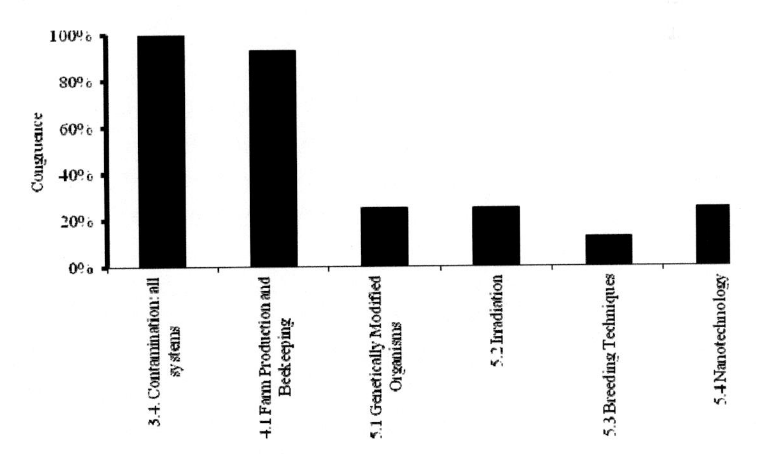

FIGURE 9.6 Congruence in some selected detailed criteria of a producer that is fully compliant with FAO's SAFA when judged against the IFOAM organic standards.

9.5 A NEED FOR INTEGRATED SUSTAINABILITY ASSESSMENTS OF ORGANIC FARMING

In 2012, the ARGOS project received funding from the NZ Government Ministry of Business, Innovation and Employment (MBIE) and several industry co-funders to develop a New Zealand Sustainability Dashboard for primary sectors (Manhire et al., 2013). This change of direction was to assist New Zealand farmers to measure and report across a rapidly expanding set of sustainability criteria incorporated into market assurance and monitoring schemes, and partly to bridge the gap between organic standards and such schemes (Figures 9.4–9.6). However, our change of emphasis was also driven by realisation that no one farming system would deliver hugely advanced sustainability or ecosystem services compared to any other. We were more struck by the large variation in sustainability outcomes between individual ARGOS farms within the same farming system panel than in relatively slight shifts in the average performance of each panel. Lifting the overall sustainability and resilience of New Zealand agriculture will depend more on assisting all farmers to do better, not from advocacy of a single farming system approach as a one-size-fits-all solution to the challenges and opportunities for future farming. Our goal was to create a practical, locally and globally relevant package of tools to turn compliance and auditing requirements into a learning opportunity for farmers and agricultural processors.

Internationally recognised frameworks and their key generic sustainability performance indicators have been co-opted to ensure that overseas consumers can benchmark and verify the sustainability credentials of New Zealand exported products. It is a participatory, industry-led approach to measuring and reporting sustainability allowing farmers to log mainly self-assessed sustainability measures into an online network. The Sustainability Dashboard will allow for instant benchmarking, trend analysis, progress towards targets and provide warnings when trigger points indicate a need for intervention. The Dashboard will also be equipped with an automated reporting system to benchmark a participating farmer's performance with that of others producing similar goods, or using similar farming technologies (eg. irrigation). The overarching framework developed in this project closely aligns with the SAFA sustainability goals and criteria but the emphasis of different parts of the assessment is adjusted to tune to New Zealand ecological, social, economic, and governance constraints and opportunities. Relatively standardised measures of farming performance will be shared between farmers, industry advocates, policy makers and consumers. A basic version of the dashboard is currently being customised and extended to meet the needs of New Zealand organic growers in particular so that organic producers can formally measure and demonstrate their performance against many of the sustainability criteria demanded by competing market assurance programmes as well as those needed for BioGro organic certification.

9.6 CONCLUSIONS: ARE ORGANIC STANDARDS SUFFICIENT TO ENSURE SUSTAINABLE AGRICULTURE?

Organic agriculture often leads to enhanced ecosystem services, as emphasised by several papers in this special journal issue (Delate et al., 2015; Abbott, 2015; Cambardella, 2015) and our selection of examples from the ARGOS project (Figure 9.2, Table 9.1). This will assist land-sharing approaches to multifunctional agriculture which can be safely assumed to promote sustainability and agricultural systems resilience. However, productivity of organic farming is often reduced compared to IM and conventional farming and this could undermine land-sparing approaches to achieving global food security while conserving ecosystem services over larger spatial scales. Some indicators of ecosystem service were relatively unchanged between farming systems, probably in part because other ecological, social, economic or governance constraints trump the effects of organic input restrictions. Provision of ecological refuges, reduced reliance on ecological subsidies, specific farming decisions like fencing shelterbelts or planting native rather than introduced trees have strong positive impacts on ecological ecosystem services, but are not part of the standards and specific requirements of organic certification. More generally, our gap analysis emphasises that organic standards only cover less than half of the broader social, economic and governance criteria for sustainability of any food and fibre production system. In contrast, farmers performing well according to accepted sustainability criteria (i.e. SAFA) would cover the majority of the organic farming requirements. Agriculture is a complex and adaptive system that responds to coupled social, ecological, economic, and governance feedbacks. It seems obvious that simple adherence to organic input restrictions and standards cannot possibly be sufficient in itself to secure sustainability and resilience. Input restrictions remain the predominant tenets of the organic standards, but wider organic principles have recently been incorporated. Current developments of the concept of Organic 3.0, which includes an attempt to demand that organic farms should demonstrate a degree of continuous development vis-à-vis the principles and goals rather than just comply with rules (IFOAM 2015), is a valuable step in this direction. We encourage strenuous promulgation of these valuable general organic principles to dispel a general and outdated notion amongst growers, policy makers and customers that organic farming is simply about restriction of certain types of potentially dangerous farm inputs. We are not advocating that organic farmers become entirely like their IM counterparts – it is vital that the organic movement retains its certified points of difference that underpin price premiums and philosophy—but we do urge organic growers to adopt the best of the IM approaches that do not compromise organic principles.

This broadening of emphasis and an organic market share defence strategy could direct best farming practice, monitoring and reporting across a wider set of sustainability criteria than simply compliance with the existing organic standards. Many of the broader criteria that are now being included in general agricultural sustainability and resilience assessments will support, and be supported by, the organic principles, even though they are not explicitly codified in the standards. Some form of 'Organics Plus' eco-verification to match the claims of green market assurance programmes could help organic growers challenge and learn from IM approaches. Each version of the New Zealand Sustainability Dashboard is hosted by a particular agricultural sector that will adjust their emphasis and investment in measuring performance to match their own particular opportunities and challenges. An organic production dashboard could therefore emphasise points of difference in organic farming methods, especially strategy to minimise risk by restricting the nature of farm inputs, while still measuring the comparative performance of organic farming on the additional dimensions demanded by other market competitors. We conclude that adherence to organic standards undoubtedly promises some gains in ecosystem services, including the crucial cultural ones that assist systems adaptability and learning—but we also assert that organic standards will need to be combined with more targeted farming systems interventions across multiple criteria to maximise sustainability of organic farming.

Until detailed measurement of the comparative performance of IM and organic farming over this wider set of criteria are tabled, it is impossible to judge whether the beneficial effects of restriction to organic inputs more than outweighs the benefits of applying a wider range of sustainability interventions while still allowing chemical inputs and similar technologies on IM farms. However, we are not advocating just another round of binary comparisons of outcomes from organic and IM farming, nor from IM and conventional farming. A safer and globally more effective approach is to find local solutions for raising ecosystem services of all farms, be they organic, IM or conventional. Systematic and targeted measurement of key agricultural ecosystem drivers, will provide feedback to enable individual farming families to locally tune their farming practices for efficient and profitable production while leaving the land fit for future generations' survival and enjoyment.

REFERENCES

1. Abbott, L. K., & Manning, D. A. C. (2015). Soil health and related ecosystem services in organic agriculture.
2. Sustainable Agriculture Research, 4(3), 116-125. http://dx.doi.org/10.5539/sar.v4n3p116
3. Bàrberi, P. (2015). Functional biodiversity in organic systems: The way forward? Sustainable Agriculture Research, 4(3), 26-31. http://dx.doi.org/10.5539/sar.v4n3p26
4. Bengtsson, J. T., Ahnstrom, J., & Weibell, A. C. (2005). The effects of organic agriculture on biodiversity and abundance: a meta-analysis. Journal of Applied Ecology, 42, 261-269. http://dx.doi.org/10.1111/j.1365-2664.2005.01005.x
5. Cambardella, C. A., Delate, K., & Jaynes, D. B. (2015). Water quality in organic systems. Sustainable Agriculture Research, 4(3), 60-69. http://dx.doi.org/10.5539/sar.v4n3p60
6. Campbell, H., Fairweather, J., Manhire, J., et al. (2012a). The Agriculture Research Group On Sustainability Programme: a longitudinal and transdisciplinary study of agricultural sustainability in New Zealand. ARGOS Research Report No. 12/01. 123 + xi pp.
7. Campbell, H., Rosin, C., Hunt, L., & Fairweather, J. (2012b). The Social Practice of Sustainable Agriculture under Audit Discipline: Initial Insights from the ARGOS Project in New Zealand. Journal of Rural Studies. 28, 129-141. http://dx.doi.org/10.1016/j.jrurstud.2011.08.003
8. Carey, P., Moller, H., Norton, S., et al. (2010). A perspective on differences in soil properties between organic and conventional farming in dairy and sheep and beef sectors. Proceedings of the New Zealand Grassland Association, 72, 35-42
9. CEO Group, (2015). Building a performance focused culture and organisation. Website: Retrieved April 1, 2015, from http://www.ceo.co.nz/performance.php
10. Darnhofer, I., Fairweather, J., & Moller, H. (2010). Assessing a farm's sustainability: Insights from resilience thinking. International Journal of Agricultural Sustainability, 8, 186-198. http://dx.doi.org/10.3763/ijas.2010.0480
11. Delate, K., Cambardella, C., Chase, C., & Turnbull, R. (2015). A review of long-term organic comparison trials in the U.S. Sustainable Agriculture Research, 4(3), 5-14. http://dx.doi.org/10.5539/sar.v4n3p5
12. Fairweather, J., Hunt, L., Benge, J., et al. (2009a). New Zealand Farmer and Orchardist Attitude and Opinion Survey 2008: Characteristics of organic, modified conventional (integrated) and organic management, and of the sheep/beef, horticulture and dairy sectors. ARGOS Research Report 09/02.
13. Fairweather, J., Hunt, L., Rosin, C., & Campbell, H., (2009b). Are conventional farmers conventional? Analysis of the environmental orientations of conventional New Zealand farmers. Rural Sociology, 74, 430-454. http://dx.doi.org/10.1526/003601109789037222

14. FAO (2013). Sustainability assessment of food and agriculture systems: SAFA Tool, beta version 2.1.50. Food and Agriculture Organization of the United Nations, Rome. Retrieved from http://www.fao.org/fileadmin/templates/nr/sustainability_pathways/docs/SAFA_Tool_User_Manual_2.1.50. pdf
15. Fukuda, Y., Moller, H., & Burns, B. (2011). Effects of organic farming, fencing and vegetation origin on spiders and beetles within shelterbelts on dairy farms. New Zealand Journal of Agricultural Research, 54, 155-176. http://dx.doi.org/10.1080/00288233.2011.591402
16. Halberg, N., Panneerselvam, P., & Treyer, S. (2015). Eco-functional intensification and food security: Synergy or compromise? Sustainable Agriculture Research, 4(3), 126-139. http://dx.doi.org/10.5539/sar.v4n3p126
17. Heckman, J. R. (2015). The role of trees and pastures in organic agriculture. Sustainable Agriculture Research, 4(3), 103-115. http://dx.doi.org/10.5539/sar.v4n3p51
18. Hole, D. G., Perkins, A. J., Wilson, J. D., Alexander, I. H., Grice, P. V., & Evans, A. D. (2005). Does organic farming benefit biodiversity? Biological Conservation, 122, 113-130. http://dx.doi.org/10.1016/j.biocon.2004.07.018
19. Hunt, L., Fairweather, J., Rosin, C., et al. (2011). Doing the unthinkable: linking farmer's breadth of view and adaptive propensity to the achievement of social, environmental and economic outcomes. Proceedings 18th International Farm Management Association Congress 'Thriving in a global market: innovation, co-operation and leadership', Methven, Canterbury, New Zealand. pp. 197-203.
20. IFOAM (International Federation of Organic Agriculture Movements). (2005). Principles of organic agriculture. http://www.ifoam.org/about_ifoam/principles/index.html
21. IFOAM (International Federation of Organic Agriculture Movements). (2015). What is Organic 3.0? http://www.ifoam.bio/en/what-organic-30
22. Köpke, U., Athmann, M., Han, E., & Kautz, T. (2015). Optimising cropping techniques for nutrient and environmental management in organic agriculture. Sustainable Agriculture Research, 4(3), 15-25. http://dx.doi.org/10.5539/sar.v4n3p15
23. Lindenmayer, D., Cunningham, S., & Young, A. (2012). Landuse Intensification. Effects on agriculture, biodiversity and ecological processes. CSIRO Publishing.
24. MacLeod, C. J., Blackwell, G., & Benge, J. (2012). Reduced pesticide toxicity and increased woody vegetation cover account for enhanced native bird densities in organic orchards. Journal of Applied Ecology, 49, 652-660. http://dx.doi.org/10.1111/j.1365-2664.2012.02135.x
25. Magbanua, F. S., Townsend, C. R., Blackwell, G. L., et al. (2010). Responses of stream macroinvertebrates and ecosystem function to conventional, integrated and organic farming. Journal of Applied Ecology, 47, 1014-1025. http://dx.doi.org/10.1111/j.1365-2664.2010.01859.x
26. Manhire, J., Moller, H., Barber, A., et al. (2012). The New Zealand Sustainability Dashboard: Unified monitoring and learning for sustainable agriculture in New Zealand. ARGOS Working Paper No. 8. 40 + vi pages. Retrieved from www.argos.org.nz
27. Meadows, S. (2012). Can birds be used as tools to inform resilient farming and environmental care in the development of biodiversity-friendly market accreditation systems? Perspectives of New Zealand sheep and beef farmers. Journal of Sustainable Agriculture, 36, 759-787. http://dx.doi.org/10.1080/10440046.2012.672375
28. Millennium Ecosystem Assessment. (2005). Ecosystems and Human Well-being: Synthesis. Island Press, Washington, DC.

29. Moller, H., MacLeod, C., Haggerty, J., et al. (2008). Intensification of New Zealand agriculture: Implications for biodiversity, New Zealand Journal of Agricultural Research 51, 253-263. http://dx.doi.org/10.1080/00288230809510453
30. Moller, H., Wearing, A., Perley, C., et al. (2007). Biodiversity on kiwifruit orchards: the importance of shelterbelts. Acta Horticulturae 753, 609-618.
31. Niggli, U. (2015). Incorporating Agroecology Into Organic Research –An Ongoing Challenge. Sustainable Agriculture Research, 4(3), 149-157. http://dx.doi.org/10.5539/sar.v4n3p149
32. Norton, S., Lucock, D., Moller, H., & Manhire, J. (2010). Energy return on investment for dairy and sheep/beef farms under conventional, integrated or organic management. Proceedings of the New Zealand Grasslands Association, 72, 145-150.
33. Pilati, A., Vanni, M. J., González, M. J., & Gaulke, A. K. (2009). Effects of agricultural subsidies of nutrients and detritus on fish and plankton of shallow-reservoir ecosystems. Ecological Applications, 19, 942-960. http://dx.doi.org/10.1890/08-0807.1
34. Pretty J., Sutherland, W. J., Ashby, J., et al. (2010) The top 100 questions of importance to the future of global agriculture, International Journal of Agricultural Sustainability, 8, 219-236. http://dx.doi.org/10.3763/ijas.2010.0534
35. Rosin, C., Perley, C., Moller, H., & Dixon, K. (2008). For want of the social, was the biodiversity battle lost? On the need to approach social-ecological resilience through transdisciplinary research. New Zealand Journal of Agricultural Research, 51, 481-484. http://dx.doi.org/10.1080/00288230809510480
36. Rozzi, P., Miglior, F., & Hand, K. (2007). A total merit selection index for Ontario organic dairy farmers. Journal of Dairy Science 90: 1584-1593. http://dx.doi.org/10.3168/jds.S0022-0302(07)71644-2
37. Sato, K., Bartlett, P., Erskine, R., & Kaneene, J. (2005). A comparison of production and management between Wisconsin organic and conventional dairy herds. Livestock Production Science, 93, 105-115. http://dx.doi.org/10.1016/j.livprodsci.2004.09.007
38. Tuuli, M. M. (2012). Competing models of how motivation, opportunity, and ability drive job performance in project teams. In. Laryea, et al., (Eds). Proceedings of the 4th West Africa Built Environment Researchers (WABER) Conference (pp. 1359 – 1366). Abuja, Nigeria.
39. Todd, J. H., Malone, L. A., McArdle, B. H., Benge, J., Poulton, J., Thorpe, S., & Beggs, J. R. (2011). Invertebrate community richness in New Zealand kiwifruit orchards under organic or integrated pest management. Agriculture, Ecosystems & Environment, 141, 32-38. http://dx.doi.org/10.1016/j.agee.2011.02.007
40. Vaarst, M. (2015). The role of animals in eco-functional intensification of organic agriculture. Sustainable Agriculture Research, 4(3), 103-115. http://dx.doi.org/10.5539/sar.v4n3p103
41. Vogl, C. R., Kummer, S., Leitgeb, F., Schunko, C., & Aigner, M. (2015). Keeping the actors in the organic system learning: The role of organic farmers' experiments. Sustainable Agriculture Research, 4(3), 140-148. http://dx.doi.org/10.5539/sar.v4n3p140

CHAPTER 10

An Ecologically Sustainable Approach to Agricultural Production Intensification: Global Perspectives and Developments

Amir Kassam and Theodor Friedrich

10.1 INTRODUCTION

Challenges arising from global economic and population growth, pervasive rural poverty, degrading natural resources in agriculture land use, and climate change are forcing ecological sustainability elements to be integrated into agricultural production intensification. The situation has been exacerbated by the fact that the quality and direction of the dominant, tillage-based, agricultural production systems world-wide, and the agricultural supply chains that support them, have moved dangerously off course onto a path of declining productivity and increasing negative externalities (MEA, 2005; WDR, 2008; McIntyre et al., 2008; Foresight, 2011). This path is considered to be unsustainable ecologically as well as economically and socially, and is being driven by the consequences of unquestioned faith and reliance on the dominant 'industrialised agriculture' mentality of technological interventions of genetics and agrochemicals in tillage-based agriculture (DEFRA, 2002, 2008; Kassam, 2008).

Now increasingly known as the 'old paradigm', this way of farming since WWII was seen as the best option for production intensification and agricultural development to keep hunger and famine at bay after WWII. Subsequently, this paradigm was thought to be a partial solution also for poverty alleviation in the developing countries.

This version of agriculture, whether industrialised or not, in which the soil structure, soil life and organic matter are mechanically destroyed every season and the soil has no organic cover, is no longer sufficiently adequate to meet the agricultural and rural resource management needs and demands of the 21st century. The future requires farming to be multi-functional and at the same time ecologically, economically and socially sustainable so that it can deliver ecosystem goods and services as well as livelihoods to producers and society. Farming needs to effectively address local, national and international challenges. These challenges include: food, water and energy insecurity, climate change, pervasive rural poverty, and degradation of natural resources.

It is now clear that the root cause of our agricultural land degradation and deceasing productivity—as seen in terms of loss of soil health—is our low soil-carbon farming paradigm of intensive tillage which disrupts and debilitates many important soil-mediated ecosystem functions. The decrease in soil carbon due to tillage occurs even more rapidly in the tropics due to higher temperatures compared with temperate zones. For the most part our soils in tillage-based farming without organic surface residue protection are becoming de-structured, our landscape is exposed and unprotected by organic mulch, and soil life is deprived of habitat and starved of organic matter. Taken together, this loss of soil biodiversity, increase in soil organic matter decease, destruction of soil structure and its biological recuperating capacity, increased soil compaction, runoff and erosion, and infestation by pests, pathogens and weeds, reflect the current degraded state of the health of many of our soils globally (Montgomery, 2007).

Further, the condition of our soils is being exacerbated by:

- applying excessive mineral fertilisers on to farm land that has been losing its ability to respond to inputs due to degradation in soil health, and
- reducing or doing away with crop diversity and rotations (which were largely in place around the time of WWII) due to agrochemical inputs and commodity-based market forces.

The situation is leading to further problems of increased threats from insect pests, diseases and weeds against which farmers are forced to apply ever more pesticides and herbicides, and which further damage biodiversity and pollute the environment.

However, we also know that the solution for sustainable farming has been known for a long time, at least since the mid-thirties when the mid-west of USA suffered massive dust storms and soil degradation due to intensive ploughing and harrowing of the prairies. For instance, in 1943, Edward Faulkner wrote a book *The Ploughman's Folly* in which he stated that there is no scientific evidence for the need to plough. More recently, David Montgomery in his well-researched book *Dirt: The Erosion of Civilizations* shows that generally with any form of tillage including non-inversion tillage the rate of soil degradation (loss of soil health) and soil erosion is greater than the rate of soil formation.

According to Montgomery's research, tillage has caused the destruction of agricultural resource base and of its productive capacity nearly everywhere throughout human history, and continues to do so. For these natural science writers as far back as 1943, tillage is not compatible with sustainable agriculture. We only have to look at the various international assessments of the large-scale degradation of our land resource base and the loss of productivity globally to reach a consensus as to whether or not the further promotion of any form of tillage-based agriculture is a wise development strategy. We contend that to continue with intensive tillage agriculture now verges on irresponsibility towards society and nature. Thus we maintain that with tillage-based agriculture in all agro-ecologies that can meet climatic, soil and terrain requirements for crop growth, no matter how different and unsuitable they may seem for no-till farming, crop productivity (efficiency) and output cannot be optimized to the full potential. There might be environments and situations, for example water logging, where no-tillage would not work satisfactorily. But if the root problems in those cases cannot be addressed differently, for example by drainage, it is questionable whether tillage based farming would be advisable under these conditions as alternative. Crop farming in such places should better be abandoned. Further, agricultural land under tillage is not fully able to deliver the needed range and quality of environmental services that are mediated by ecosystem functions in the soil system. Obviously, something must change.

10.2 FARMING PARADIGMS

Essentially, we have two farming paradigms operating, and both aspire to sustainability. (1) The tillage-based farming systems, including intensive tillage with inversion ploughing during the last century, aims at modifying soil structure to create a clean seed bed for planting seeds and to bury weeds or incorporate residues. This is the interventionist paradigm in which most aspects of crop production are controlled by technological interventions such as soil tilling, modern varieties, protective or curative pest, pathogen and weed control with agrochemicals, and the application of mineral fertilizers for plant nutrition. (2) The no-tillage farming systems, since the forties or so, allow for a predominantly ecosystem approach, and can be productive and ecologically sustainable. This is the agro-ecological paradigm characterised by minimal disturbance of the soil and the natural environment, use of traditional or modern adapted varieties, plant nutrition from a mix of organic and non-organic sources including biological nitrogen fixation, an integrated approach to pest management leaving curative pesticides as a last resort, and the use of both natural and managed biodiversity to produce food, raw materials and other ecosystem services. Crop production based on an ecosystem or agro-ecological approach can sustain the health of farmland already in use, and can regenerate land left in poor condition by past misuse.

The post-WWII agricultural policy placed increasing reliance upon 'new' high yielding seeds, more intensive tillage of various types and heavy and more powerful machines, combined with even more chemical fertilizers, pesticides and herbicides, and mono-cropping. According to our reading, factories producing nitrates for manufacturing explosives needed for WWII quickly had to find an alternate market once the war ended. The crop production sector was a sitting target for the explosives salesmen who went around convincing farmers that high yields and more profit could be obtained with mineral nitrogen and that there was presumably no real need for crop diversification and rotations with legumes or for adding organic sources of plant nutrients or animal manure. This technological interventionist approach became the accepted paradigm for production intensification, and was promoted globally along with genetically enhanced modern varieties–referred to as the Green Revolution paradigm of the '50s and '60s that included the Asian Green Revolution particularly in the irrigated rice-wheat systems in the Indo-Gangetic Plains of south Asia. While the Green Revolution raised crop yields and total output of food staples, and averted a looming food crisis in south Asia, it also resulted in the following situation in most agricultural landscapes in the tropics and outside in the sub-tropics and temperate environments:

- loss of SOM (soil organic matter), porosity, aeration, biota (=decline in soil health) -> collapse of soil structure -> surface sealing, often accompanied by mechanical compaction, -> decrease in infiltration-> waterlogging -> flooding) (Figure Fig. 10.1);
- loss of water as runoff and of soil as sediment;
- loss of time, seeds, fertilizer, pesticide (erosion, leaching);
- less capacity to capture and slowly release water and nutrients;
- less efficiency of mineral fertilizer: "The crops have become 'addicted' to fertilizers";
- loss of biodiversity in the ecosystem, below & above soil surface;
- more pest problems (breakdown of food-webs for micro-organisms and natural pest control);
- falling input efficiency & factor productivities, declining yields;
- reduced resilience, reduced sustainability;
- poor adaptability to climate-change and its mitigation; and
- higher production costs, lower farm productivity and profit, degraded ecosystem services.

FIGURE 10.1 Consequence of intensive-tillage paradigm. Notice that due to soil compaction and loss in water infiltration ability caused by regular soil tillage leads to impeded drainage and flooding after a thunder storm in the ploughed field (right) and no flooding in the no-till field (left). Photograph taken in June 2004 in a plot from a long-term field trial "Oberacker" at Zollikofen close to Berne, Switzerland, started in 1994 by SWISS NO-TILL. The three water filled "cavities" in the no-till field derive from soil samples taken for "spade tests" prior to the thunder storm.
(Credit: Wolfgang Sturny)

10.3 A SOLUTION: NO-TILL AGRO-ECOLOGICAL SYSTEM

Conservation Agriculture (CA), also known as a 'no-till' farming system, is an effective solution to stopping agricultural land degradation, for rehabilitation, and for sustainable crop production intensification. CA has gained momentum in North and South America, in Australia and New Zealand, in Asia in Kazakhstan and China, and in the southern Africa region.

CA has the following three core inter-linked principles (Friedrich et al., 2009):

- No or minimum mechanical soil disturbance and seeding or planting directly into undisturbed or untilled soil, in order to maintain or improve soil organic matter content, soil structure and overall soil health.
- Enhancing and maintaining organic mulch cover on the soil surface, using crops, cover crops or crop residues. This protects the soil surface, conserves water and nutrients, promotes soil biological activity and contributes to integrated weed and pest management.
- Diversification of species—both annuals and perennials—in associations, sequences and rotations that can include trees, shrubs, pastures and crops, all contributing to enhanced crop and livestock nutrition and improved system resilience.

These principles and key practices appear to offer an entirely appropriate alternative to most modern and traditional tillage-based agricultural production systems in the tropical, sub-tropical and temperate agro-ecologies, with a potential capacity to slow and reverse productivity losses and environmental damages. In conjunction with other complementary good crop management practices for integrated crop nutrition, pest and water management, and good quality adapted seeds, the implementation of the CA principles provide a solid foundation for sustainable production intensification.

These principles can be integrated into most rain-fed and irrigated production systems to strengthen their ecological sustainability, including horticulture, agro-forestry, organic farming, System of Rice Intensification (SRI), 'slash and mulch' rotational farming, and integrated crop-livestock systems, CA is a lead example of the agro-ecological paradigm for sustainable production intensification now adopted by FAO as seen in its recent publication 'Save and Grow'. Empirical evidence shows that farmer-led transformation of agricultural production systems based on Conservation Agriculture (CA) is gathering momentum globally. CA, comprising minimum mechanical soil disturbance (no-till and direct seeding), organic soil cover, and crop species diversification, is now estimated to be practiced globally on about 125 M ha (some 9% of global arable cropland) across all continents (Table 10.1) and all agricultural ecologies, with some 50% of the area located in the developing regions. During the last decade, cropland under CA has been increasing yearly at a rate of some 7 million hectares, mainly in the Americas, Australia, and, more recently, in Asia and Africa (Friedrich et al., 2012).

TABLE 10.1 Area under CA by continent

Continent	Area (hectare)	Percent of total
South America	55,464,100	45
North America	39,981,000	32
Australia & New Zealand	17,162,000	14
Asia	4,723,000	4
Russia & Ukraine	5,100,000	3
Europe	1,351,900	1
Africa	1,012,840	1
World total	**124,794,840**	**100**

For the farmer the initial drivers for adoption of CA are mostly erosion or drought problems, as well as cost pressure. However, drivers of change that are valid for large-scale farmers are different from small-scale farmers. Water erosion has been the main driver in Brazil, wind erosion and cost of production in the Canadian and American Prairies, and drought and cost issues in Australia and Kazakhstan. More recently, concern about the economic and environmental unsustainability of traditional approaches to agriculture internationally, including small-scale farming in Africa and Asia, has stimulated governments to seriously consider CA whose principles can be implemented by small or large farmers in most agro-ecologies to raise productivity and harness environmental services, avoid and recuperate from land degradation, and respond to climate change.

In the adoption of or transformation to CA, there are constraints and opportunities that must be addressed in different ways in different places depending on their nature. These include:

- Weeds that can be controlled using integrated management practices involving a combination of surface mulch, cover crops, rotations, mechanical management and herbicides.
- Net labour requirement which by and large is reduced over time with increase in labour productivity (in terms of output per unit input) in all CA systems whether with manual, animal or mechanised farm power.
- Larger farmers are not the only beneficiaries of CA. Small farmers with any farm power source can practice CA and harness a range of benefits. Similarly, field-based horticulture production can also benefit from CA, whether small or large scale.
- Livestock can create a competition for residues but over time CA generates more biomass which can permit effective management of functional biomass to meet the needs of livestock and of soil health. A combination of on-farm livestock management and area integration of crop-livestock with community participation provides a basis for overcoming this constraint.
- Temperate areas of Europe are claimed to be different from other areas where CA has been widely adopted.

This appears to be a myth, as seen from the viewpoint of almost a 'wholesale' transformation to CA in some states in Canada and in Western Australia and parts of USA, and more recently the introduction and growing evidence of CA in Finland, Switzerland, UK, France, Italy, Germany and Denmark. Constraints to CA adoption appear to be surmountable for up-scaling when:

- Farmers are working together in testing and sharing experience and generating new knowledge, and using the innovation network approach as an effective way of CA extension.
- Appropriate and affordable no-till equipment and machinery is available.
- There is relevant and problem solving knowledge generation and technical capacity in the research and extension system to offer advice to farmers, industry and policy makers.
- Eventual risks involved in transforming to no-till systems are buffered through appropriate insurances and/ or incentives.
- There is effective policy and institutional support for adoption and widespread uptake.

In the developing regions, especially among larger mechanized farms there has been spontaneous adoption of CA. However, the adoption process more generally, including for small-holder farmers, is still slow and has not yet entered into the exponential uptake phase. In recent years the situation has begun to change in Asia and Africa and there is already growing government commitment and programmes in these regions to promote CA, including for small scale farmers. In Africa, the Southern Africa region is at an advanced stage of early adoption with countries such as South Africa, Zambia, Zimbabwe, Mozambique, Malawi and Tanzania leading the way. In Asia, small farmers in China and India and large farmers in Kazakhstan have made significant progress with no-till systems in recent years.

However, a more coordinated approach and harmonized policy will be needed for CA to really take off and provide benefits to small holder farmers and bring land degradation under control. Empirical evidence across many countries has shown that the rapid adoption and spread of CA requires a change in behaviour of all stakeholders. For the farmers, a mechanism to experiment, learn and adapt is a prerequisite.

Policy-makers and institutional leaders need to fully understand the longer-term productivity, economic, social and environmental benefits of CA for producers and the society at large.

10.4 CA–AN OPPORTUNITY TO SAVE AND MAKE MONEY, ALLEVIATE RURAL POVERTY INTERNATIONALLY AND TO IMPROVE THE PLANET

Advantages offered by CA to small or large farmers include better livelihood and income. For the small farmer under a manual system, CA offers ultimately 50% labour saving, less drudgery, more stable yields, and improved food security. To the mechanised farmers CA offers lower fuel use and less machinery and maintenance costs. Reduced cost of production with CA is a key to better profitability and competitiveness, as well as keeping food affordable.

Against the background of rising input, food and energy costs, land degradation and climate change, experience of switching to CA confirms that the known advantages include higher soil carbon levels, microorganism and meso fauna activity over time, minimisation or avoidance of soil erosion, the reversal of soil degradation, improved aquifer recharge due to greater density and depth of soil biopores due to more earthworms and more extensive and deeper rooting. CA advantages also include adaptation to climate change due to increased infiltration and soil moisture storage and increased availability of soil moisture to crops, reduced runoff and flooding, and improved drought and heat tolerance by crops, and climate change mitigation through reduced emissions due to 50-70% lower fuel use, 20-50% lower fertilizer/pesticides, 50% reduction in machinery and use of smaller machines, C-sequestration of 0.20-0.7 or more t.ha-1.y-1 depending on the ecology and residue management, and no excess CO_2 release as a result of no burning of residues (Kassam et al., 2009; Corsi et al., 2012).

To the community and society, CA offers public goods that include: less pollution, lower cost for water treatment, more stable river flows with reduced flooding and maintenance, and cleaner air and less siltation of dams (Mello and Raij, 2006; ITAIPU, 2011). At the landscape level, CA offers the advantages of better ecosystem services including: provision of food and clean water, regulation of climate and pests/diseases, support of nutrient cycles, pollination, cultural recreation, enhancement of biodiversity, and erosion control. At the global level, the public goods are: improvements in groundwater resources, soil resources, biodiversity and mitigation of climate change (Haugen-Kozyra and Goddard, 2009).

CA is highly relevant to several elements of the global agenda. It is the base element for combining intensive, highly productive agriculture with sustainability and ecosystem services, which responds to the strategic Objective A of FAO that deals with the promotion of sustainable production intensification based on a new paradigm of agriculture (FAO, 2011), improving the prospects for achieving the UN Millennium Development Goals. For the future of agriculture, CA comprises the best available set of agro-ecological concepts and production practices for climate change adaptation and mitigation, addressing the risks of climate variability, and reducing the vulnerability to drought, flood, heat, frost and wind. There is worldwide evidence from research and farmer practice to show that large productivity, economic, social and environmental benefits for the farmers and for the society can be harnessed through the adoption of CA practices.

For example, if agriculture is to provide a significant sink for carbon and to drastically reduce greenhouse gas emissions, this can be done cost-effectively through wide scale adoption of CA, thus contributing to climate change mitigation. Further, CA helps to improve rural livelihoods by contributing to rural poverty reduction and eventually even to reversing the rural-urban migration trends.

However, there is still a need for more concerted and sustained efforts to promote CA globally, requiring the involvement of all sectors and stakeholders, from farmers, research and extension across to input supply industries and output value chain service providers to policy makers and institutional leaders to educational and vocational training institutions. It is this integrated stakeholder engagement that has been responsible for the rapid uptake of CA in large parts of the Americas and Australia.

10.5 PRO-POOR PRODUCTION SYSTEMS AND POVERTY ALLEVIATION

Currently, it is estimated that three-quarter of the bottom billion are rural-based and rely on agriculture for their food security and livelihood. As long as they 'must' remain in agriculture as producers and agriculture workers, every effort should be made to help producers to adopt sustainable production systems such as CA (FAO, 2011, www.fao.org/ag/ca), and System of Rice Intensification (SRI) (Kassam et al., 2011b, http://sri.ciifad.cornell.edu) which are effective in pro-poor development for small farmers, as well as have the potential for enabling small farmers to produce 'more from less' and can offer surplus food to the local markets at a lower price. SRI is an alternative agro-ecological approach to rice production in which the soil is not flooded but kept moist, allowing the rice plant to grow a large root system. Together with a different set of crop management practices including transplanting young seedlings in wider spacing, SRI methods lead to increased yield and reduction in the use of production inputs of seeds, water, nutrients, pesticides and even labour.

With aerobic soil conditions, SRI system can be integrated into CA systems, offering further productivity and environmental benefits, including reduced methane emission. A quarter of the bottom billion is urban-based. Their food security will depend on wage employment within the economy, as it grows and diversifies, to be able to purchase affordable food from the market. Any safety-net social support for urban-based as well as rural-based poor families, in terms of cash and access to food rations, would help to improve food security. Similarly, training of youth and adult from the urban and rural poor families for skills development would improve chances of wage employment. Support to educate the children of poor urban and rural families would eventually help them to break out of the downward spirals and poverty traps.

However, in the long run, all stakeholders–farmers, supply and value chain service providers, academics, researchers, extension agents, policy makers, civil servants, consumers–must become engaged in understanding and harnessing the full power of the no-till agro-ecological paradigm. This will contribute to making farming and rural resource management careers an attractive source of livelihood to future generations who must take a custodial view of their role in managing the planet's natural resources for food security and economic development.

10.6 CONCLUDING REMARKS

CA principles appear to be universally applicable because the practice of CA is not a blanket recommendation or recipe for everywhere (also called silver bullet or "panacea") but has to be adapted to the site and farmer circumstances. CA produces more from less, can be adopted and practiced by smallholder poor farmers, builds on the farmer's own natural resource base, does not entirely depend on purchased derived inputs, and is relatively less costly even in the early stages of sustainable production intensification. More emphasis should be put on the constraints and challenges in overcoming the hindrances in tropical and subtropical small-scale farmer areas in Africa and Asia and the solutions that might be different from the larger scale farmers of Brazil and Argentina.

CA being a new paradigm for most farmers globally, special emphasis must be placed on the need of a change in mindset amongst farmers especially in traditional farming communities in the North and the South and the importance of involving all stakeholders to apply a holistic approach in CA promotion that is just as much farmer driven as it is science driven and supported by public and private sectors and national agriculture development policies.

REFERENCES

1. DEFRA (2002) Farming and Food: A Sustainable Future. Report of the Policy Commission on the Future of Farming and Food. January 2002. DEFRA, UK
2. DEFRA (2009c) Safeguarding Our Soils: A strategy for England. September 2009. DEFRA,UK (http://www.defra.gov.uk/environment/quality/land/soil/documents/soil-strategy.pdf)
3. FAO (2011). Save and Grow: a policymaker's guide to the sustainable intensification of smallholder crop production. FAO, Rome, 102 pp.
4. Foresight (2011). The Future of Food and Farming. The Government Office for Science, London.
5. Friedrich, T., Kassam, A. H. and Shaxson, F. (2009). Conservation Agriculture. In: Agriculture for Developing Countries. Science and Technology Options Assessment (STOA) Project. European Parliament. European Technology Assessment Group, Karlsruhe, Germany.

6. Friedric, T., Derpsch, R. and Kassam, A. H. (2012). Global overview of the spread of Conservation Agriculture. Journal Agriculture Science and Technology (in press).
7. Haugen-Kozyra, K., Goddard, T. 2009. Conservation agriculture protocols for green house gas offsets in a working carbon markets. Paper presented at the IV World Congress on Conservation Agriculture, 3-7 February 2009, New Delhi, India.
8. ITAIPU (2011). Cultivando Agua Boa (Growing Good Water), (http://www2.itaipu.gov.br/cultivandoaguaboa/)
9. Kassam, A. H. (2008). Sustainability of farming in Europe: Is there a role for Conservation Agriculture? Journal of Farm Management 13 (10): 717-728.
10. Kassam, A. H., Friedrich, T., Shaxson, F. and Pretty, J. (2009). The spread of Conservation Agriculture: Justification, sustainability and uptake. Int. J. Agric. Sustainability, 7(4): 292-320.
11. Kassam, A. H., Friedrich, T. and Derpsch, R. (2010). Conservation Agriculture in the 21st Century: A Paradigm of Sustainable Agriculture. In: The Proceedings of the European Congress on Conservation Agriculture, 6-8 October 2010, Madrid, Spain.
12. Kassam, A. H., Friedrich, T., Shaxson, F., Reeves, T., Pretty, J. and de Moraes Sa, J. C. (2011a). Production Systems for Sustainable Intensification–Integrating Productivity with Ecosystem Services. Technology Assessment–Theory and Prexis, Special Issue on Feeding the World, July 2011.
13. Kassam, A. H., N. Uphoff and W. A. Stoop (2011b). Review of SRI modifications in rice crop and water management and research issues for making further improvements in agricultural and water productivity. Paddy and Water Environment 9 :1 (DOI 10.1007/s10333-011-0259-1)
14. McIntyre, B. D., Herren, H. R., Wakhungu, J. and Watson, R. T. (eds) (2008). Agriculture at a Crossroads: Synthesis. Report of the International Assessment of Agricultural Knowledge, Science, and Technology for Development (IAASTD). Washington, DC: Island Press.
15. MEA (2005). Ecosystems and Human Well-Being: Synthesis. Millennium Ecosystem Assessment. Washington, DC: Island Press.
16. Mello, I., van Raij, B. (2006). No-till for sustainable agriculture in Brazil. Proc. World Assoc. Soil and Water Conserv., P1: 49-57.
17. Montgomery, D. (2007). Dirt: the erosion of civilizations. University California Press, Berkeley and Los Angeles. 287 pp.
18. WDR (2008). Agriculture for Development. World Development Report. Washington, DC: World Bank. AK/TF: 4-3-2012

PART III

Annotated Bibliographies for Organic and Sustainable Agriculture

CHAPTER 11

Tracing the Evolution of Organic/Sustainable Agriculture: A Selected and Annotated Bibliography

Mary V. Gold and Jane Potter Gates

INTRODUCTION

> "Are we going to protect our springs of prosperity, our raw material of industry and commerce and employer of capital and labor combined; or are we going to dissipate them? According as we accept or ignore our responsibility as trustees of the nation's welfare, our children and our children's children for uncounted generations will call us blessed, or will lay their suffering at our doors." Gifford Pinchot, 1908[1]

The U.S. Department of Agriculture's National Agricultural Library (NAL) in Beltsville, Maryland, holds a vast archive of historical documentation covering all aspects of agriculture. This bibliography focuses on works and authors selected from the Library collection that pertain to sustainability in agriculture. It was compiled in the hope of increasing recognition of and access to knowledge that might help address today's challenges to a sustainable agriculture.

Mary V. Gold and Jane Potter Gates. Tracing the evolution of organic/sustainable agriculture: a selected and annotated bibliography. Beltsville, MD: United States Dept. of Agriculture, National Agricultural Library [1988]; updated and expanded, May 2007, http://afsic.nal.usda.gov/tracing-evolution-organic-sustainable-agriculture-1, August 2015.

This publication builds on an original 1988 bibliography authored by former Alternative Farming Systems Information Center (AFSIC) coordinator, Jane Potter Gates.

The idea of sustainability holds ambiguities and nuances that are sometimes difficult to resolve. Sustainable agriculture was addressed by the U.S. Congress in the 1990 "Farm Bill." Under that law,

> "the term sustainable agriculture means an integrated system of plant and animal production practices having a site-specific application that will, over the long term:
>
> - satisfy human food and fiber needs
> - enhance environmental quality and the natural resource base upon which the agricultural economy depends
> - make the most efficient use of nonrenewable resources and on-farm resources and integrate, where appropriate, natural biological cycles and controls
> - sustain the economic viability of farm operations
> - enhance the quality of life for farmers and society as a whole."[2]

"Doing" sustainable agriculture is a less complicated endeavor than defining it. People from all walks of life understand that there are practical steps that can be taken to protect the ecological and human resources that a viable food production system relies on. Many farmers in the U.S. have adopted practices commonly accepted as sustainable. "Hailing from small vegetable farms, cattle ranches and grain farms covering thousands of acres, [producers] have embraced new approaches to agriculture. They are renewing profits, enhancing environmental stewardship and improving the lives of their families as well as their communities." The New American Farmer, 2005.[3]

Organic farming holds a special place under the sustainable agriculture umbrella. The U.S. Department of Agriculture now defines standards for organic practices and for food labeled as "organic."[4] The commercial impacts of this phenomenon have added new issues to the sustainability discussion.[5]

Although sustainable and organic approaches to food production may seem relatively new, they rest on a base of science and philosophy that has been centuries in the making. The term "sustainable agriculture" did not come into popular use until the late 1980s;[6] however, we know that the notion of land stewardship is a very old one. Some of the earliest known writings reveal sophisticated stewardship ethics and practices. Archaeologists have unearthed indicators of soil and water conservation efforts from civilizations on every continent.

Historical evidence traces an ebb and flow of concern for stewardship and long-term food production over the years. As social, economic and environmental conditions evolved, so did the issues impacting sustainable use of resources. It is not surprising that particularly difficult times and places spawned the most dramatic "learning curves" in terms of both successful and failed practices and systems.

The farmers, researchers, social thinkers, educators, historians, policy makers, artists and everyday citizens represented here analyzed and proposed remedies for problems of their own eras. Some of them were "movers and shakers;" others remained obscure, to be discovered by later generations. Most researched and wrote on the edges of the "conventional wisdom" of their day. Many of them studied history themselves, looking backward for information and direction.

If we listen, voices of these forebears do several things for us. They teach us practical lessons about problems and problem solving. They provide an historical context for understanding contemporary challenges. And they inspire us with their passion. Most importantly, they remind us that history is a continuum. History describes where we have been, defines the aspirations and limitations of our current endeavors, and carries us into the future.

Challenges to a sustainable, global food system that will carry us through the coming years and into the next century are daunting. However, we have access to a storehouse of tools with which to work: a diverse agricultural knowledgebase; interdisciplinary research and expertise; cutting-edge technology applications; and a global communication system with which to share information.

> "Our television documentaries and books show us in graphic detail why the Easter Islanders, Classic Maya and other past societies collapsed. Thus, we have the opportunity to learn from mistakes of distant peoples and past peoples. That's an opportunity that no past society enjoyed to such a degree." Jared Diamond, 2006[7]

REFERENCES THROUGH 1799

The environmental impacts of early agriculture remain undocumented, at least in terms of written records. Archaeological discoveries provide clues. For instance, rock paintings found in the Tibesti Mountains of North Africa dating to 3,500 BCE indicate that overgrazing and desertification were phenomena even in ancient times. The pictures show that areas of the Sahara described as barren in the first century A.D. were once fertile fields.

On varying timetables, people in Europe, Asia, Africa and the Americas developed agricultural systems, many incorporating land stewardship practices. Examples include: the Chinese adoption of composting and mulching (ca. 2000 BCE); Sumerian irrigation and windbreak techniques evidenced in document tablets from circa 1500 BCE; and the Aztec Chinampas (floating garden) system utilized since the 1100s that produces several corn crops per year. (From: The People's Chronology: A Year-by Year Record of Human Events from Prehistory to the Present, by James Trager, New York: Henry Holt, 1994) Several works written during the 19th and 20th centuries that document these and other ancient systems are cited in this bibliography. See: Dickson, 1788; King, 1911; Wrench, 1946; Lowdermilk, 1948; Osborn, 1948; Hyams, 1952; and Diamond, 2005.

0050

Columella, Lucius Junius Moderatus, 6-70

De Re Rustica

Cambridge MA: Harvard University Press, 1941 (oldest edition held by the National Agricultural Library). 3 vols. Loeb Classical Library edition. Translated title: On Agriculture. Recension of the text and an English translation by Harrison Boyd Ash. Bibliography, v. 1, p. xxiii-xxvii. Other editions: The National Agricultural Library holds various editions of this work, published in Italy, England and Germany, dating from the 1500s.

NAL Call no: 30.8 C72Ag

Full-text: Columella: Extant Works (De Re Rustica and De Arboribus), Bill Thayer's Web Site, http://penelope.uchicago.edu/Thayer/E/Roman/Texts/Columella/home.html (accessed Jan. 1, 2007)

Annotation: In the first century, Columella wrote, "The earth neither grows old, nor wears out, if it be dunged." He also recommended grains in rotation with legumes and fallow. Cato, Varro, Palladius, Vegetius and Pliny the Elder also wrote about soil building and conservation techniques. MVG

1580

Tusser, Thomas, 1524?-1580

Five Hundred Points of Good Husbandry

London: H. Denham, 1580. 4p., 289 numbered leaves, 2p. Complete title: Fiue hundred pointes of good husbandrie, as well for the champion, or open countrie, as also for the woodland, or seuerall, mixed in eurie month with huswiferie, ouer and besides the booke of huswiferie, corrected, better ordered, and newly augmented to a fourth part more, with diuers other lessons, as a diet for the farmer, or the properties of winds, planets, hope, herbes, bees, and approoued remedies for sheepe and cattle, with many other matters both profitable and not vnpleasant for the reader: also a table of husbandrie at the beginning of this book: and another of huswiforie at the end: for the better and easier finding of any matter contained in the same. In verse.

NAL Call no: 30.8 T87 1580

Annotation: This classic has been reprinted almost every century since its original publication. Tusser's maxims include observations of human behavior - "Still crop' upon crop many farmers do take and reap little profit, for greediness sake... 11, observations concerning the land" and "land (overburdened) is clean out of heart," or "if land be unlusty, the crop is not great." He also gives advice by the month, frequently in rhyme. "Octobers Abstract" is about the rotation of crops: "Where barlie did growe, laie wheat to sowe, yet better I thinke, sowe pease, after drinke. And then if ye please, sowe wheat after pease." JPG

Cited in: Bailey (1915)

1748

Eliot, Jared, 1685-1763

Essays Upon Field-husbandry in New-England, as It is or May be Ordered

Boston: Edes and Gill, 1760. 166p. "The foregoing essays were first printed in New-London and in New-York; the 1st in 1748, 2d in 1749, 3d in 1751, 4th in 1753, 5th in 1754, 6th in 1759." p. 158. Appendix dated June, 1761. Other editions: Essays upon Field Husbandry in New England, and Other Papers, 1748-1762, edited by Harry J. Carman and Rexford G. Tugwell, with a biographical sketch by Rodney H. True (Columbia University Press, 1934).

NAL Call no: 31.3 El4E R

Annotation: Eliot was a minister, doctor, philosopher, author and scientist-farmer. His six essays, based on observations and experiments made at his farm in Connecticut, were the first American publications devoted to agriculture. He adapted English practices, recommending legume/grain rotations and control of erosion on hillsides including his own brand of conservation tillage. "When our fore-Fathers settled here, they entered a Land which probably never had been Ploughed since the Creation; the Land being new they depended upon the natural Fertility of the Ground, which served their purpose very well and when they had worn out one piece they cleared another, without any concern to amend their Land, except a little helped by the Fold and Cart-dung, whereas in England they would think a Man a bad Husband, if he should pretend to sow Wheat on Land without any Dressing." MVG

Cited in: McDonald (1941)

1788

Dickson, Adam, 1721-1776

The Husbandry of the Ancients

Edinburgh: Dickson and Creeca, 1788. 2 vols.: 527p. and 494p.

NAL Call no: R30.9 D56

Other works by this author: A Treatise on Agriculture (1762, later editions, 1765, 1785); Small Harms Destructive to the Country in its Present Situation (1764); Essay on Manures (1772).

Annotation: Dickson quotes Columella, Palladius, Cato, Virgil, Pliny, et. al., regarding the knowledge and practice of husbandry and confesses in the preface to be "agreeably surprised to find, that, not-withstanding the great differences in climate, the maxims of the ancient Roman farmers are the same with those of the best modern farmers in Britain..." JPG

Cited in: Pieters (1927)

1790

Deane, Samuel, 1733-1814

The New-England Farmer, or, Georgical Dictionary: Containing a Compendious Account of the Ways and Methods in Which the Important Art of Husbandry, in All its Various Branches, is, or May be Practised, to the Greatest Advantage, in this Country

Worcester MA: Isaiah Thomas, 1790. 335p.

NAL Call no: 30.1 D34 Ed.1

Full-text: (1797 edition) Internet Archive, http://www.archive.org/details/newenglandfarmer00deanrich (accessed Jan. 1, 2007)

Annotation: Deane was the first American to document the problem of wind erosion. Along with other soil-conserving practices such as green manuring and contour plowing, he recommended windbreaks and hedgerows to prevent "sand-floods." His original 1790 book was updated several times and was reportedly a mainstay of New England farmers until the Civil War. MVG

Cited in: McDonald (1941)

1800–1899

THE AGE OF DISCOVERY'S AGRICULTURAL LEGACY

Scientific and geographical discoveries of the 1500s, 1600s and 1700s had great impact on agriculture and on the use of natural resources worldwide. Significant changes included: accelerated "globalization" of plant and animal species; unprecedented advances in scientific knowledge and research techniques; access to new markets and trading partners; and social revolutions that redefined labor and land ownership.

For the Europeans, undeveloped continents seemed to offer a never-ending supply of arable land and cheap labor. This had special significance for the Americas. "The felling of the first tree by colonists in the New World, though never mentioned by historians, was an act of great significance. It marked the beginning of the era of the most rapid rate of wasteful land use in the history of the world." Early American Soil Conservationists, by Angus McDonald. Washington DC: United States Department of Agriculture, 1941.

1804

Saussure, Nicolas-Théodore de, 1767-1845

Récherches Chimiques sur la Végétation

Paris: Chez la Ve. Nyon, 1804. viii, 327p. In French. Translated title: Chemical Research on Vegetation.

NAL Call no: 463.2 .S285R

Annotation: The majority of this Swiss scientist's papers deal with the chemistry and physiology of plants and plant interactions with soils. His work clarified many previously misunderstood soil-plant relationships including plant respiration and the soil's role as supplier of nitrogen. These findings set the scene for modern soil science and humus-oriented theory. MVG

Cited in: Korcak (1992)

1813

Taylor, John, 1753-1824

Arator: Being a Series of Agricultural Essays, Practical and Political, in Sixty-one Numbers, by a Citizen of Virginia

Georgetown, DC: J.M. and J.B. Carter, 1813. 296p. Other editions: Several later editions of this work appeared in the early 1800s; a recent volume, edited and with an introduction by M.E. Bradford was published in 1977.

NAL Call no: 30 T21

Other works by this author: *An Inquiry into the Principles and Policy of the Government of the United States* (1814).

Annotation: Taylor was a statesman, author, Virginia plantation owner and close friend of Thomas Jefferson. His observations of soil depletion led to his investigation and advocacy of soil restorative practices. His widely read essays and newspaper articles recommended the following: protect the soil from grazing during the rest period, and thus raise a large crop of vegetable matter; make use of vegetable manures of all kinds; sow clover and grass seed with the grain crop to serve as pasturage or green-manure; practice horizontal plowing as a preventative of gullies and washes; and establish artificial meadows and a crop rotation with grass. MVG

1832

Ruffin, Edmund, 1794-1865

An Essay on Calcareous Manures

Petersburg VA: J. W. Campbell, 1832. xii, 13, 242p. Other editions: Several subsequent editions published through 1852.

NAL Call no: 57.1 R83 R

Other works by this author: *An Address on the Opposite Results of Exhausting and Fertilizing Systems of Agriculture: Read before the South-Carolina Institute, at its*

Fourth Annual Fair, November 18th, 1852 (1853); *Agricultural, Geological, and Descriptive Sketches of Lower North Carolina, and the Similar Adjacent Lands* (1861); *Agriculture, Geology, and Society in Antebellum South Carolina: The Private Diary of Edmund Ruffin*, 1843, edited by William M. Mathew (1992).

Annotation: Ruffin's work and writings cover many topics related to sustaining farm production in Virginia and North Carolina, including lime-soil interactions (especially "marling"), flood and sedimentation control, cover crops, soil exhaustion and wind erosion. "Edmund Ruffin's efforts ended the pioneer state of the erosion-control movement in America. His work was equal to that of all his predecessors combined. The knowledge of the soil which he gained from his experiments, his theories and speculations regarding the action of water on soil and his erosion-control practices provided a foundation for later developments." A. McDonald, Early American Soil Conservationists (1941), p. 58. MVG

Cited in: McDonald (1941)

1840

Liebig, Justus Freiherr von, 1803-1873

Organic Chemistry and its Application to Agriculture and Physiology

London: Taylor and Walton, 1840. 407p. Edited from the manuscript of the author by Lyon Playfair. Translated from the German, Organische Chemie in ihrer Anwendung auf Agricultur und Physiologie. Includes bibliographical references.

NAL Call no: 395 L62O

Other works by this author: *Chemical Letters* (2nd corrected edition) (full-text: Soil and Health Library, Steve Solomon, http://www.soilandhealth.org/01aglibrary/01principles.html) (accessed Apr. 23, 2007).

Annotation: Liebig's work established basic chemical requirements for agricultural production and plant nutrition. His discoveries, coupled with those of Sir Humphrey Davy (Elements of Agricultural Chemistry, 1813) reduced the soil-plant relationship to chemical reactions and an agricultural "revolution" was begun. By the 1940s, large-scale use of synthetic chemical fertilizers had become mainstream. Liebig's legacy marks the divergent paths of "conventional" and organic agriculture. MVG

Cited in: Conford (2001); Kirschenmann (2004); Korcak (1992)

1842

Dana, Samuel Luther, 1795-1868
A Muck Manual for Farmers
Lowell MA: Daniel Bixby, 1842. 242p.
NAL Call no: 56 D19 1842
Other works by this author: *Manures: A Prize Essay* (1844); *Essay on Manures* (1850).
Annotation: "A treatise on the physical and chemical properties of soils; the chemistry of manures; including also the subjects of composts, artificial manures and irrigation." (From the title of the 5th edition of this work, published 1855.) One of the first American-published books to elaborate on the science of soil improvement, this work includes information about the use of city-generated organic wastes and industrial by-products as agricultural soil amendments. MVG
Cited in: Blum (1993)

1846

Allen, Richard Lamb, 1803-1869

A Brief Compend of American Agriculture

New York: Saxon and Miles, 1846. 437p. Includes index.

NAL Call no: 31.3 AL5 1846

Other works by this author: *Domestic Animals. History and Description of the Horse, Mule, Cattle, Sheep, Swine, Poultry, and Farm Dogs. With Directions for their Management, Breeding, Crossing, Rearing, Feeding, and Preparation for a Profitable Market. Also, their Diseases and Remedies Together with Full Directions for the Management of the Dairy* (1847) (full-text: Core Historical Literature of Agriculture; http://chla.library.cornell.edu/cgi/t/text/text-idx?c=chla;idno=3058099) (accessed Apr. 23, 2007); New American Farm Book, with Lewis F. Allen (1858) (full-text: Making of America Books, http://quod.lib.umich.edu/cgi/t/text/text-idx?c=moa;idno=AJR0646.0001.001) (accessed Apr. 23, 2007).

Annotation: The introduction contains a proposal for establishment of a "National Board of Agriculture," plus recommendations for States' actions, particularly regarding education. JPG

Cited in: Pieters (1927)

1853

Browne, Daniel Jay, 1804-1867?

The Field Book of Manures, or, the American Muck Book: Treating of the Nature, Properties, Sources, History, and Operations of All the Principal Fertilisers and Manures in Common Use, with Specific Directions for Their Preparation, Preservation, and Application to the Soil and to Crops; as Combined with the Leading Principles of Practical and Scientific Agriculture; Drawn from Authentic Sources, Actual Experience, and Personal Observation

New York: C.M. Saxton and Company, Agricultural Book Publishers, 1855 (oldest edition held by the National Agricultural Library). xii, 5, 422p. Illustrated with engravings.

NAL Call no: 57 B81

Full-text: Internet Archive, http://www.archive.org/details/fieldbookofmanur00browuoft (accessed Jan. 1, 2007)

Other works by this author: *Sylva Americana or a Description of the Forest Trees Indigenous to the United States* (1832); *The American Poultry Yard; Comprising the Origin, History, and Description of the Different Breeds of Domestic Poultry, with Samuel Allen* (1850) (full-text: Internet Archive, http://www.archive.org/details/americanpoultryy00browrich) (accessed Apr. 23, 2007).

Annotation: As the book title indicates, Browne presents information pertaining to soil productivity through incorporation of organic matter. He served as head of the agricultural division of the Patent Office from June 9, 1853, through 1859. He also published a book (cited above) calling for native tree planting directed especially at farmers. MVG

Cited in: Harwood (1983); Harwood (1990); Kirschenmann (2004)

1854

Thoreau, Henry David, 1817-1862

Walden; or, Life in the Woods

Garden City NY: Doubleday, 1960. 280p. Reprint of the first edition of Walden, published in Boston by Ticknor and Fields in 1854.

NAL Call no: 145 T39

Full-text: The Thoreau Reader, Iowa State University and the Thoreau Society, http://thoreau.eserver.org/walden00.html (accessed Jan. 1, 2007)

Annotation: Thoreau's advocacy for living simply, in accordance with nature, included a strong conservation message for agriculture. "By avarice and selfishness and a grovelling habit, from which none of us is free, of regarding the soil as property, or the means of acquiring property chiefly, the landscape is deformed, husbandry is degraded with us and the farmer leads the meanest of lives. He knows Nature but as a robber." Chapter 7. MVG

Cited in: Esbjornson (1992)

1860

Sorsby, Nicholas T.

Horizontal Plowing and Hill-side Ditching

Mobile: S.H. Goetzel, 1860. 45p.

NAL Call no: n.a.

Annotation: Sorsby was a Southern planter who farmed in Mississippi and Alabama. His book on tillage and erosion management for farmers advocates a system of level, contour tillage patterns; ridge and furrow plowing; exact grading methods; gully rehabilitation and drainage ditches. His system was complicated and ahead of its time, but its principles were often cited in soil conservation efforts of later decades. MVG

Cited in: McDonald (1941)

1864

Marsh, George Perkins, 1801-1882

Man and Nature, or, Physical Geography as Modified by Human Action

New York: C. Scribner, 1869 (oldest edition held by the National Agricultural Library). xix, 577p. "Bibliographical list of works consulted" p. vii-xv. Other editions: The Earth as Modified by Human Action: A Last revision of "Man and Nature," 1907; John Harvard Library edition, 1965.

NAL Call no: 331 M35E 1869

Full-text: Making of America Books, http://www.hti.umich.edu/cgi/t/text/text-idx?c=moa;idno=AJA7231.0001.001 (accessed Jan. 1, 2007)

Other works by this author: Irrigation: Its Evils, the Remedies, and the Compensations (1873); So Great a Vision: The Conservation Writings of George Perkins Marsh, edited by Stephen C. Trombulak (2001).

Annotation: Influential lawyer, diplomat and scholar, Marsh, was one of the first to recognize and describe in detail the significance of human action in transforming the natural world and to advocate society's responsibility in addressing it. Piqued by the damage farmers in his native Vermont did by clear-cutting their land, he broadened his scope to study and discuss ecological problems on an international scale. This book, first published in 1864, has become a classic of environmental literature. MVG

1865

Wolfinger, John F.

Green Manuring and Manures

In Report of the U.S. Commissioner of Agriculture for the Year 1864. Washington DC: Government Printing Office (1865), p. 299-328.

NAL Call no: 1 Ag84 1864

Full-text: National Agricultural Library Digital Repository (NALDR), http://naldr.nal.usda.gov/NALWeb/Search.aspx (search on "green manuring" and scroll to "Report of the U.S. Commissioner of Agriculture for the Year 1864;" click on page numbers 223-270 in table) (accessed Jan. 1, 2007)

Annotation: Lincoln was President when this report was published. Wolfinger defines his subject, tracing its history in Flanders (now Belgium) and listing the benefits of and objections to, the practice of green manuring. He also quotes what "the best agricultural writers say of green manures." JPG

Cited in: Pieters (1927)

1878

Riley, Charles Valentine, 1843-1895

The Rocky Mountain Locust: Its Metamorphoses and Natural Enemies

Publisher unknown: 1878. 1 vol.

NAL Call no: 429 R45Ro

Annotation: Riley was a scientist, artist and prolific writer who is credited with helping establish the field of modern entomology. "One of Riley's greatest triumphs while Chief of the Federal Entomological Service (1881-1894) was his initiation of efforts to collect parasites and predators of the cottony cushion scale, which was destroying the citrus industry in California. In 1888, he sent Albert

Koebele to Australia to collect natural enemies of the scale. A beetle, Vedalia cardinalis, now Rodolia cardinalis, was introduced into California and significantly reduced populations of the cottony cushion scale. This effort gave great impetus to the study of biological control for the reduction of injurious pests and established Charles Valentine Riley as the 'Father of the Biological Control.'" Charles Valentine Riley Collection, Biographical Notes, National Agricultural Library (http://specialcollections.nal.usda.gov/guide-collections/charles-valentine-riley-papers (link is external)) (accessed Apr. 23, 2005). MVG

1881

Darwin, Charles Robert, 1809-1882

The Formation of Vegetable Mould Through the Action of Worms with Observations on their Habits

London: Murray, 1945. Reprint of the 1881 publication.

NAL Call no: 56.12 D45

Full-text: Soil and Health Library, Steve Solomon, http://www.soilandhealth.org/01aglibrary/01principles.html (accessed Jan. 1, 2007)

Annotation: The real foundation for "the study of the principles underlying farming and gardening..." JPG

Cited in: Coleman (1976); Harwood (1983); Harwood (1990); Kirschenmann (2004); Korcak (1992); Merrill (1983)

1894

Hensel, Julius

Bread from Stones. A New and Rational System of Land Fertilization and Physical Regeneration

Philadelphia PA: A. J. Tafel, 1894. 140p. Translated from the German. Other editions: 2nd edition, 1911; 3rd edition, 1913. Reissued, Acres U.S.A., 1991.

NAL Call no: 57.6 B74

Annotation: Using his understanding of the role of earth minerals in the production of food crops, Hensel championed the use of "stonemeal," as natural fertilzer and recommended excluding animal manures and commercial fertilizers from farming. J.I. Rodale, in his book, Pay Dirt, notes that this theory was first offered by Hensel in Norway in 1885. Although few growers today espouse excluding organic soil amendments, rock dust and rock minerals are recognized as essential ingredients in soil building practices. MVG

1897

Roberts, Isaac Phillips, 1833-1928

The Fertility of the Land: A Summary Sketch of the Relationship of Farm-practice to the Maintaining and Increasing of the Productivity of the Soil

New York: Macmillan, 1897. xvii, 415p. Preface by L. H. Bailey. Other editions: Several revised versions published through 1909. (Rural Science Series)

NAL Call no: 57 R54

Full-text: Core Historical Literature of Agriculture, Cornell University, http://chla.library.cornell.edu/cgi/t/text/text-idx?c=chla;idno=2846741 (accessed Jan. 1, 2007)

Other works by this author: The Production and Care of Farm Manures (1891); Soil Depletion in Respect to the Care of Fruit Trees (1895); Ten Acres Enough; A Practical Experience Showing How a Very Small Farm May be Made to Keep a Very Large Family, with Edmund Morris (1905); Autobiography of a Farm Boy (1946).

Annotation: Roberts was both a farmer and an agricultural educator. His observation that, "If land contains a reasonable amount of potential plant-food and fails to give satisfactory results, it would appear to be both unbusinesslike and unscientific to add plant-food rather than to use that already in possession," was fundamental to his philosophy and understanding of land productivity. MVG

Cited in: Harwood (1983); Harwood (1990)

1898

Bentley, Henry Lewis

Cattle Ranges of the Southwest: A History of the Exhaustion of the Pasturage and Suggestions for its Restoration

Washington DC: U.S. Dept. of Agriculture, 1898. 32p. (Farmers' Bulletin, No. 72)

NAL Call no: 1 Ag84F no.72

Full-text: Organic Roots, Organic Agriculture Information Access, http://www.hti.umich.edu/n/nal/ (to be added June, 2007) (accessed Jan. 1, 2007)

Other works by this author: A Report upon the Grasses and Forage Plants of Central Texas (1898); Experiments in Range Improvement in Central Texas (1902).

Annotation: Bentley describes the early condition of central Texas ranges and the factors that contributed to their deterioration during the late 1800s. His recommendations on how the value of the stock ranges could be renewed through appropriate stocking rates, water conservation practices, hay production and the use of native grasses and forage plants seem remarkably contemporary. MVG

1898

Frank, Albert Bernhard, 1839-1900

A Manual of Agricultural Botany

Edinburgh; London: W. Blackwood and Sons, 1898. x, 199p. Translated from the German, Pflanzenkunde für Mittlere und Niedere Landwirthschaftschülen (1894), by John Waugh Paterson. Illustrated with 133 woodcuts.

NAL Call no: 64 F85

Annotation: Frank is best known for his research on plant-fungi symbiosis related to truffle production. He is credited with inventing the term, "mycorrhiza," in the paper, "On the Nourishment of Trees through a Root Symbiosis with Underground Fungi" (1885), Proceedings of the German Botanical Association (full-text in German: http://www.biologie.uni-hamburg.de/b-online/fo33/frank/frank.htm) (accessed Apr. 23, 2007). MVG

Cited in: Merrill (1983)

1900–1944

THE RISE OF "SCIENTIFIC AGRICULTURE" AND ADOPTION OF MANUFACTURED CHEMICAL FERTILIZERS AND PESTICIDES

In the years after Liebig's revelations about soil chemistry and plant nutrition (see: Liebig, 1840), most farmers and agricultural researchers adopted chemically-oriented soil and crop management techniques that they saw as more scientific than traditional practices. The large-scale use of synthetic fertilizers came slowly, but surely. It was coupled, in the years following World War I and World

War II, with the use of newly developed chemicals that were used to control insect pests and weeds.

Simultaneously, this shift to chemically- and technologically-intensive farming was accompanied by attitudinal and scientific changes that helped shape modern organic and sustainable agriculture. These included: the study and acceptance of "biological control" techniques; renewed interest in the role of humus and soil microorganisms in plant production; and innovative approaches to composting. There were also many scientific discoveries concerning human nutrition and the relationship among agricultural practices, food and human diseases.

During this period Americans were confronted with evidence of deteriorated of rangelands, soils and forests. The first critics of the new "industrial" agriculture emerged and a heightened conservation ethic began to take root. The New Zealand Soil and Health Association (originally called the "Humic Compost Club") was founded in 1942, pre-dating both the British Soil Association (1946) and Rodale's Soil and Health Foundation (1947).

1906

Hilgard, Eugene Woldemar, 1833-1916

Soils, their Formation, Properties, and Relations to Climate and Plant Growth in the Humid and Arid Regions

New York: Macmillian, 1906. 593p. Includes index. List of authors referred to in text.

NAL Call no: 56 H54S

Full-text: Core Historical Literature of Agriculture, Cornell University, http://chla.library.cornell.edu/cgi/t/text/text-idx?c=chla;idno=2845620 (accessed Jan. 1, 2007)

This author published many other works on agricultural topics under the auspices of the Berkeley CA Agricultural Experiment Station including: *The "Bedrock Lands" of Sacramento County,* with R.H. Loughridge (1885); *The Distribution of the Salts in Alkali Soils,* with R.H. Loughridge (1895); *The Conservation of Soil Moisture and Economy in the Use of Irrigation Water,* with R.H. Loughridge (1898).

Annotation: Professor of Agriculture at the University of California and Director of the California Experiment Station, Hilgard originally planned this to be a text and reference book, but enlarged its scope to include his soil studies "in the humid and arid regions." JPG

Cited in: Merrill (1983); Pieters (1927)

1907

Elliot, Robert Henry, 1837-1914

The Clifton Park System of Farming and Laying Down Land to Grass

London: Simpkin, et.al., 1907. 260p. Includes index. First published in 1898 as Agricultural Changes. Introduction by Sir George Stapledon.

NAL Call no: 32 E152 (not listed in catalog, 1/2007)

Full-text: (5th edition) Soil and Health Library, Steve Solomon; and Journey to Forever Online Library, http://www.soilandhealth.org/01aglibrary/01principles.html; http://journeytoforever.org/farm_library.html (accessed Jan. 1, 2007)

Annotation: The author writes of his more than 30 years' experience in India and in England as a planter and a farmer. He quotes Cato, devotes considerable space to Arthur Young and remarks that proposals to agricultural changes are often met with a response characterized as "What we knows we knows, and what we don't know we don't want to know." JPG

Cited in: Coleman (1976); Conford (1988); Harwood (1983); Harwood (1990)

1907

Fletcher, Stevenson Whitcomb, 1875-1971

Soils: How to Handle and Improve Them

New York: Doubleday, Page, 1907. 438p. Includes index. Appendix (includes crop rotation recommendations by state). Illustrated with photographs by the author. (The Farm Library)

NAL Call no: 56.7 F632

Full-text: Core Historical Literature of Agriculture, Cornell University, http://chla.library.cornell.edu/cgi/t/text/text-idx?c=chla;idno=2716150 (accessed Jan. 1, 2007)

Other works by this author: *The Strawberry in North America: History, Origin, Botany, and Breeding* (1917); *Pennsylvania Agriculture and Country Life, 1840-1940* (1955).

Annotation: Fletcher, then at the Agricultural College of Michigan, attempts here to "set forth the important facts about the soil in a plain and non-technical manner." JPG

Cited in: Harwood (1983); Harwood (1990)

1908

Pinchot, Gifford, 1865-1946

The Conservation of Natural Resources

Washington DC: U.S. Dept. of Agriculture, 1908. 12p. (Farmers' Bulletin, 327)

NAL Call no: 1 Ag84F no.327

Full-text: Organic Roots, Organic Agriculture Information Access, http://www.hti.umich.edu/n/nal/ (to be added June, 2007) (accessed Jan. 1, 2007)

Other works by this author: The Fight for Conservation (1910); The Power Monopoly: Its Make-up and its Menace (1928); Breaking New Ground.(1947); The Conservation Diaries of Gifford Pinchot, edited by Harold K. Steen (2001).

Annotation: "We shall decide whether their [our children] lives, on the average, are to be lived in a flourishing country, full of all that helps to make men comfortable, happy, and strong, and effective, or whether their lives are to be lived in a country like the miserable outworn regions of the earth which other nations before us have possessed without foresight and turned in to hopeless deserts. We are no more exempt from the operation of natural laws than are the people of any other part of the world." p. 12. Pinchot addresses not only the status and future of forest and soil resources in the United States, but "Waste through piecemeal planning" and the "Danger of monopoly." MVG

1910

Carver, George Washington, 1864?-1943

George Washington Carver in his Own Words

Columbia MO: University of Missouri Press, 1987. xv, 208p. Edited by Gary R. Kremer. Includes index. Bibliography, p. 197-205.

NAL Call no: S417.C3C3

Annotation: Lacking funds for research at Tuskegee Institute, Carver worked on improving soils, growing crops with few inputs and using species that fixed nitrogen (hence, the work on the cowpea and the peanut). He emphasized providing information that farmers needed, presented at the level they could use. "This pegs him as one of the first true sustainable agriculture educators and researchers." Dennis Keeney, in Leopold Letter, Fall 1998 MVG

1910

Hopkins, Cyril George, 1866-1919

Soil Fertility and Permanent Agriculture

Boston: Ginn, 1910. 653p. Includes index. Appendix. See also: History, and Present Position of the Rothamsted Investigations, by Sir J. Henry Gilbert (Harrison, 1891) (full-text: Internet Archive, http://www.archive.org/details/historypresentpo00gilbuoft) (accessed Apr. 23, 2007). (Country Life Education Series)

NAL Call no: 56.6 H77

Full-text: Core Historical Literature of Agriculture, Cornell University, http://chla.library.cornell.edu/cgi/t/text/text-idx?c=chla;idno=3057996 (accessed Jan. 1, 2007)

Other works by this author: *The Story of Soil* (1910) (full-text: Soil and Health Library, Steve Solomon, http://www.soilandhealth.org/01aglibrary/01principles.html) (accessed Apr. 23, 2007); *The Farm that Won't Wear Out* (1913) (full-text: Soil and Health Library, Steve Solomon, http://www.soilandhealth.org/01aglibrary/01principles.html) (accessed Apr. 23, 2007).

Annotation: This book contains chapters on "Theories concerning soil fertility" and on the Rothamsted Experiments. The final chapter is titled "Two Periods in Agricultural History" and contains quotations from Varro (B.C.226 to 28) to Liebig and Lincoln (1859) to King (1910). JPG

Cited in: Beeman (1993); Merrill (1983); quoted in King (1911)

1911

King, Franklin Hiram, 1848-1911

Farmers of Forty Centuries (or) Permanent Agriculture in China, Korea, and Japan

Madison WI: Mrs. F. H. King, 1911. 379p. Includes index. 209 illustrations. Preface by L.H. Bailey. Other editions: 2nd edition, edited by J.P. Bruce, London, 1927.

NAL Call no: 34.5 K58

Full-text: Core Historical Literature of Agriculture, Cornell University, http://chla.library.cornell.edu/cgi/t/text/text-idx?c=chla;idno=2917542 (accessed Jan. 1, 2007)

Other works by this author: *Destructive Effects of Winds on Sandy Soils and Light Sandy Loams: With Methods of Protection* (1894); *The Soil: Its Nature, Relations, and Fundamental Principles of Management* (1895); *Irrigation and Drainage: Principles and Practice of their Cultural Phases* (1899).

Annotation: King, a chief of the USDA Division of Soil Management, wrote this book after his retirement but did not live to write a final chapter. Bailey calls it the "writing of a well-trained observer who went to study the actual conditions of life of agricultural peoples." It is one of the most influential of all the works cited, with far-reaching consequences for agricultural practices worldwide. JPG

Cited in: Beeman (1993); Coleman (1976); Conford (2001); Esbjornson (1992); Harwood (1983); Harwood (1990); Heckman (2006); Kirschenmann (2004); Korcak (1992); Merrill (1983); Pieters (1927)

1911

United States Country Life Commission

Report on the Commission on Country Life

New York: Sturgis and Walton, 1911. 150p. Report first printed as 60th Congress, 2nd session, Senate document 705 (1909). Introduction by Theodore Roosevelt. Commission members: L.H. Bailey, Henry Wallace, Kenyon L. Butterfield, Walter H. Page, Gifford Pinchot, C.S. Barrett and W.A Beard. Other editions: Reprint published by Arno, 1975.

NAL Call no: 281.2 Un32 1911

Full-text: Core Historical Literature of Agriculture, Cornell University (1909 edition), http://chla.library.cornell.edu/cgi/t/text/text-idx?c=chla;idno=3319041 (accessed Jan. 1, 2007)

Annotation: Healthy rural economies and communities are key components of a sustainable agriculture. This report, commissioned by Theodore Roosevelt, describes "corrective forces" that should be implemented to address "deficiencies in country life" including disregard for farmer and farm laborer rights, land speculation, waste in forests, soil depletion, transportation, health, women's work and trade restraints. "We were founded as a nation of farmers and in spite of the great growth of our industrial life it still remains true that our whole system rests upon the farm, that the welfare of the whole community depends on the farmer. The strengthening of country life is the strengthening of the whole nation." Introduction, T. Roosevelt. MVG

1913

Sampson, Arthur William, 1884-1967

Range Improvement by Deferred and Rotation Grazing

Washington DC: U.S. Dept. of Agriculture, 1913. 16p. (Bulletin of the U.S. Department of Agriculture. No. 34)

NAL Call no: 1 Ag84B no.34

Other works by this author: *Reseeding of Depleted Grazing Lands to Cultivated Forage Plants* (1913); *Range Preservation and its Relation to Erosion Control on Western Grazing Lands,* with Leon H. Weyl (1918); *Climate and Plant Growth in Certain Vegetative Associations* (1918) (full-text: Internet Archive, http://www.archive.org/details/climateplantgrow00samprich) (accessed Apr. 1, 2007); *Plant Succession in Relation to Range Management* (1919); *Range and Pasture Management* (1923); *Grazing Periods and Forage Production on the National Forests,* with Harry E. Malmsten (1926); *Range Management: Principles and Practices* (1952).

Annotation: "Arthur William Sampson's list of 'firsts' is impressive: first person in America to be called a range ecologist, first to promote deferred and rotational grazing strategies, first to develop usable concepts of indicator species and plant succession for evaluating range condition, first to write a college text on range management, first range ecologist hired by the Forest Service, and first director of what is now called the Intermountain Research Station." Utah History to Go (http://historytogo.utah.gov/people/utahns_of_achievement/arthurwilliamsampson.html) (accessed Apr. 1, 2007) MVG

1915

Bailey, Liberty Hyde, 1858-1954

The Holy Earth

New York: Scribner's, 1915. 117p.

NAL Call no: 30.4 B15

Full-text: Library of Congress, American Memory, The Evolution of the Conservation Movement, 1850-1920, http://lcweb2.loc.gov/ammem/amrvhtml/conshome.html (accessed Jan. 1, 2007)

Other works by this author: *Cyclopedia of American Agriculture: A Popular Survey of Agricultural Conditions, Practices and Ideals in the United States and Canada* (4-volumes, 1907-1909) (full-text: Core Historical Literature of Agriculture Library, http://chla.library.cornell.edu/c/chla/browse/title/2949859.html) (accessed Apr. 23, 2007).

Annotation: "Dean" Bailey, who wrote many textbooks in fields relating to agriculture and who published one volume of verse, writes here on a simple "philosophy of rural life." JPG

Cited in: Kirschenmann (2004); Merrill (1983); quoted in Bromfield (1947)

1915

United States Department of Agriculture

Social and Labor Needs of Farm Women

Washington DC: United States Department of Agriculture, 1915. (Report, 103)

NAL Call no: 1Ag848p no.103

Full-text: History Matters, http://historymatters.gmu.edu/d/101/ (accessed Jan. 1, 2007)

Related USDA reports: *Domestic Needs of Farm Women*, Report 104 (1915); *Educational Needs of Farm Women*, Report 105 (1915); *Economic Needs of Farm Women*, Report 106 (1915).

Annotation: This pioneering report, based on excerpts of letters from farm wives, provided important insights into farm and rural life, and the roles of men and women, that had not been studied before. It was published with three accompanying reports, cited above. MVG

1919

Albrecht, William Albert, 1888-1974

The Albrecht Papers

Kansas City MO: Acres U.S.A., 1982. 2 vols. Edited by Charles Walters Jr. Contains ~70 papers in 2 volumes: v. 1. Foundation concepts; v. 2. Soil fertility and animal health. Includes indexes and Albrecht Bibliography, p. 3-37.

NAL Call no: S441.A44 1982

Full-text: Several papers as well as a collection of journal and magazine articles, experiment station and other government publications, are available at the Soil and Health Library, Steve Solomon, http://www.soilandhealth.org/01aglibrary/01principles.html (accessed Jan. 1, 2007)

Other works by this author: *Artificial Manure Production on the Farm* (1927); *Legume Bacteria with Reference to Light and Longevity*, with Lloyd M. Turk (1929); *Nitrate Production in Soils as Influenced by Cropping and Soil Treatments* (1938); *Nitrogen Fixation and Soil Fertility Exhaustion by Soybeans under Different Levels of Potassium*, with Carl E. Ferguson (1941); *Soil Fertility and Animal Health* (1958).

Annotation: Dr. Albrecht's observations, research and teaching on soil and soil's relationship to plant, animal and human nutrition reflect the essence of sustainable soil management. A professor at the University of Missouri College of Agriculture, Albrecht wrote hundreds of reports, books and articles that span several decades, starting with his reports on nitrogen fixation and soil inoculation in 1919. MVG

Cited in: Coleman (1976); Merrill (1983)

1923

Gray, L. C., O. E. Baker, F. J. Marschner, B. O. Weitz, William Ridgely Chapline, Ward Shepard and Raphael Zon

The Utilization of Our Lands for Crops, Pasture and Forests

In Agriculture Yearbook, 1923. Washington DC: United States Department of Agriculture; Government Printing Office (1923), p. 415-506.

NAL Call no: 1 Ag844 1923

Full-text: National Agricultural Library Digital Repository (NALDR), http://naldr.nal.usda.gov/NALWeb/Publications.aspx (select "Yearbook..."; then search by author name and 1923) (accessed Jan. 1, 2007)

Annotation: This article reflects the government's views as to the "present situation and future outlook" regarding available resources for the growing of food and raw materials which must be supplied by crop lands, pastures and forests. As such, it is both a summary and an estimate. JPG

Cited in: Pieters (1927)

1924

Steiner, Rudolf, 1861-1925

Agriculture: A Course of Eight Lectures

London: Bio-Dynamic Agricultural Association, Rudolf Steiner House, 1974 (oldest edition held by the National Agricultural Library). 175p. Translated from the German, Geisteswissenschaftliche Grundlagen zum Gedeihen der Landwirtschaft (1924), by George Adams. Preface by Ehrenfried Pfeiffer.

NAL Call no: S523.S8313 1974

Other works by this author: *Readings in Goethean Science*, with Johann Wolfgang von Goethe (1978); *What is Biodynamics? A Way to Heal and Revitalize the Earth: Seven Lectures* (2005).

Annotation: This is the text which is based on the series of lectures Steiner gave in Koberwitz, Silesia in 1924. The lecture series marked the beginning of the biodynamic agriculture movement. JPG. [The biodynamic farm/food certification organization, Demeter, was also initiated in the 1920s. MVG]

Cited in: Conford (2001); Harwood (1983); Harwood (1990); Heckman (2006); Kirschenmann (2004); Scofield (1986)

1927

Elton, Charles Sutherland, 1900-1991

Animal Ecology

London: Sidgwick and Jackson, 1927. 207p. Introduction by Julian S. Huxley. List of references, p.192-200. Other editions: 3rd edition, 1950.

NAL Call no: 411 E18

Other works by this author: *Animal Ecology and Evolution* (1930); *The Ecology of Invasions by Animals and Plants* (1958).

Annotation: Elton helped define the field of ecology by focusing on the study of animal populations as opposed to studying individual organisms or the more general "scientific natural history." His approach, along with refinements established by Amyan Macfadyen (Animal Ecology: Aims and Methods, 1957) and others in the 1950s, laid the groundwork for "agroecology." MVG. See also: Altieri, 1983 and Gliessman, 1998.

1927

Pieters, Adrian John, 1866-1950

Green Manuring: Principles and Practice

New York: Wiley, 1927. xiv, 356p. Includes index. Bibliography. (The Wiley Agricultural Series)

NAL Call no: 57.5 P61

Full-text: Soil and Health Library, Steve Solomon, http://www.soilandhealth.org/01aglibrary/01principles.html(accessed Jan. 1, 2007)

Other works by this author: *Green Manuring*, with C.V. Piper (USDA Farmers' Bulletin, no. 1250, 1922); *Soil-depleting, Soil-conserving, and Soil-building Crops* (USDA Leaflet, no. 165, 1938); *Legumes in Soil Conservation Practices* (USDA Leaflet, no. 163, 1938).

Annotation: Pieters was an agronomist working for the USDA at the time he wrote this book, defining "green manuring," and cover, catch and shade crops. The second chapter, a history of the subject, covers China and Japan, Greece and Rome, through the Middle Ages to England and America in the 19th Century. JPG

Cited in: Korcak (1992); Waksman (1936)

1928

Bennett, Hugh Hammond, 1881-1960; and William Ridgely Chapline

Soil Erosion: A National Menace

Washington DC: United States Department of Agriculture, 1928. 36p. 16 plates of photographs. Bibliography, p. 35-36. (Circular, 33)

NAL Call no: 1 Ag84C no.33

Full-text: Organic Roots, Organic Agriculture Information Access, http://www.hti.umich.edu/n/nal/ (to be added June, 2007) (accessed Jan. 1, 2007)

Other works by this author: *Conservation Farming Practices and Flood Control* (1936); *Soil Conservation* (1939); *Thomas Jefferson, Soil Conservationist* (1944);

Our American Land: The Story of its Abuse and its Conservation (1946); *Elements of Soil Conservation* (1947).

Annotation: "Hugh Hammond Bennett led the soil conservation movement in the United States in the 1920s and 1930s, urged the nation to address the 'national menace' of soil erosion and created a new federal agency and served as its first chief - the Soil Conservation Service, now the Natural Resources Conservation Service [NRCS] in the U.S. Department of Agriculture. He is considered today to be the father of soil conservation." A Story of Land and People: Biography of Hugh Hammond Bennett, USDA, NRCS (http://www.nrcs.usda.gov/about/history/bennett.html) (accessed February 1, 2007). MVG

Cited in: Beeman (1993); Worster (1985)

1929

Smith, Joseph Russell, 1874-1966

Tree Crops: A Permanent Agriculture

New York: Harcourt, Brace, 1929. xii, 333p. "List of articles in which the tree crops idea has been broadcasted," p. 295. "Bibliography on soil erosion and its prevention," p. 296-301. Other editions: 1950 edition (Devon-Adair) includes an introduction by Wendell Berry.

NAL Call no: 99 Sm6 1929

Full-text: Journey to Forever Online Library (selected chapters only), http://journeytoforever.org/farm_library.html (accessed Jan. 1, 2007)

Annotation: Smith's compilation of important trees and how to grow them was written to promote remedies for worn out soils, soil erosion on hillsides, flooding and degradation of arid lands. "Testing applied to the plant kingdom would show that the natural engines of food production for hill lands are not corn and other grasses, but trees." Part I: "The Philosophy, Tree Crops - The Way Out." MVG

1930

Jenny, Hans, 1899-1992

A Study on the Influence of Climate upon the Nitrogen and Organic Matter Content of the Soil

Columbia MO: University of Missouri, College of Agriculture, Agricultural Experiment Station, 1930. 66p. Bibliography, p. 64-66. (Research Bulletin, 152)

NAL Call no: 100 M693 (3) no.152

Other works by this author: *Factors of Soil Formation: A System of Quantitative Pedology* (1941, reprinted 1994) (full-text: Soil and Health Library, Steve Solomon, http://www.soilandhealth.org/01aglibrary/01principles.html) (accessed Apr. 23, 2007); *The Soil Resource: Origin and Behavior* (1980).

Annotation: Jenny's work, first at the University of Missouri (Columbia) and later at the University of California, Berkeley, focused on defining soil properties and the process of soil formation. His research on the effect of natural influences—climate, organisms, topography, time and parent material—on soil organic matter (SOM) dynamics was very influential during the mid-1900s and seems particularly relevant today, MVG

Cited in: Worster (1985)

1932

Pottenger, Francis Marion, 1901-1967

Pottenger's Cats: A Study in Nutrition

La Mesa CA: Price-Pottenger Nutrition Foundation, 1995 (oldest edition held by the National Agricultural Library). xv, 123p. Bibliography: "The professional papers of Francis M. Pottenger, Jr., M.D.," p. 121-123. Edited by Elaine Pottenger with Robert T. Pottenger, Jr.

NAL Call no: TX537 .P67 1995

Annotation: This compilation presents the observations made by Francis M. Pottenger, Jr., M.D., on the effects of deficient and optimum nutrition in cats and human beings as recorded in his articles and clinical records written between the years of 1932 and 1956. Pottenger's work focused on the nutritive value of heat-labile elements—nutrients destroyed by heat and available only in raw foods. He linked his observations of cats on deficient diets to Dr. Weston Price's studies (see Price, 1939) of human degeneration found in tribes and villages that had abandoned traditional foods. MVG

Cited in: Merrill (1983)

1935

Sears, Paul Bigelow, 1891-1990

Deserts on the March

Norman: University of Oklahoma Press, 1935. 231p. Most recent edition: 4th edition, University of Oklahoma Press, 1980.

NAL Call no: 277.12 Se1

Other works by this author: *Life and Environment* (1939); *Biology of the Living Landscape* (1964); *Lands Beyond the Forest* (1969).

Annotation: In 1935, drought held sway over much of the United States and the Dust Bowl was at its worst. With an ecological approach, Sears writes eloquently about desertification, a problem that remains one of the primary challenges facing sustainability of agriculture world-wide. MVG

Cited in: Conford (2001)

1935

Stapledon, Sir Reginald George, 1882-1960

The Land Now and Tomorrow

London: Faber and Faber, 1935. xvii, 336p. Includes foldout maps. Bibliography, p. 316-325.

NAL Call no: 282 St2

Other works by this author: *The Cultivation and Varieties of Oats* (1923); *Ley Farming*, with W. Davies (1941) (full-text, 1948 edition: Journey to Forever Online Library, http://journeytoforever.org/farm_library.html) (accessed Apr. 23, 2007); *The Way of the Land* (1949) (full-text: Soil and Health Library, Steve Solomon, http://www.soilandhealth.org/01aglibrary/01principles.html) (accessed Apr. 23, 2007).

Annotation: Stapledon, Professor of Agricultural Botany at University College of Wales Aberystwyth, founded the Welsh Plant Breeding Station in 1919. He was concerned with ley farming and improvement of grass plants. MVG

Cited in: Conford (2001)

1936

McCarrison, Sir Robert, 1878-1960

Nutrition and Health, being the Cantor Lectures Delivered before the Royal Society of Arts, 1936, together with two earlier essays, by Sir Robert McCarrison, and a postscript by H. M. Sinclair

London: Faber and Faber, 1953 (oldest edition held by the National Agricultural Library). 125p. Published in 1936 under title: Nutrition and National Health. (Royal Society of Arts, London, Cantor Lectures)

NAL Call no: 389.1 M125N 1953

Full-text: (McCarrison lectures only) McCarrison Society Scottish Group, http://www.foodforhealthscotland.org/nutritionandnationalhealth.html (accessed Jan. 1, 2007)

Other works by this author: *Studies in Deficiency Disease* (1921) (full-text: Soil and Health Library, Steve Solomon, http://www.soilandhealth.org/01aglibrary/01principles.html) (accessed Apr. 23, 2007); *The Work of Sir Robert McCarrison*, edited by H.M. Sinclair (1953).

Annotation: Dr. McCarrison spent most his career in the Indian Medical Service. Through his studies in India and Britain, dating from the 1930s, he concluded that the relationship of food to nutrition and of both to health and disease was the key to maintaining good "national health." MVG

Cited in: Conford (1988); Conford (2001); Merrill (1983)

1936

Waksman, Selman Abraham, 1888-1973
Humus: Origin, Chemical Composition, and Importance in Nature
Baltimore: Williams and Wilkins, 1936. 526p. Subject and author indexes. Extensive bibliography.
NAL Call no: 56 W13H
Full-text: Core Historical Literature of Agriculture, Cornell University, http://chla.library.cornell.edu/cgi/t/text/text-idx?c=chla;idno=2828925 (accessed Jan. 1, 2007)
Other works by this author: *Principles of Soil Microbiology* (1927); *Soil Microbiology (1952); Soil Organic Matter and the Living Plant,* with A.W. Blair (1938); *Guide to the Classification and Identification of the Actinomycetes and their Antibiotics,* with Hubert A. Lechevalier (1953); *My Life with the Microbes* (1954).
Annotation: An "attempt to tell the story of humus, its origin from plant and animal residues, its chemical composition, its physical properties, its importance in nature, especially in soil processes and in plant growth and finally its decomposition." JPG
Cited in: Balfour (1943); Beeman (1993); Blum (1993); Coleman (1976); Conford (2001); Harwood (1983); Harwood (1990); Korcak (1992); Merrill (1983); author cited extensively by Pfeiffer

1938

Pfeiffer, Ehrenfried, 1899-1961

Bio-dynamic Farming and Gardening: Soil Fertility Renewal and Preservation

New York; London: Anthroposophic Press; Rudoff Steiner Pub. Co., 1938. vii, 2 leaves, 220p. Translated from the German, Die Fruchtbarkeit der Erde (1937), by Fred Heckel. Bibliography, p. 217-220. 18 illustrations.

NAL Call no: 30 P47

Full-text: (American edition) Core Historical Literature of Agriculture, Cornell University, http://chla.library.cornell.edu/cgi/t/text/text-idx?c=chla;idno=3138977 (accessed Jan. 1, 2007)

Other works by this author: *Practical Guide to the Use of the Bio-dynamic Preparations* (revised edition, 1945); *The Earth's Face and Human Destiny* (1947); *The Compost Manufacturers Manual; the Practice of Large Scale Composting* (1956); *Weeds and What They Tell* (reprint, 1981).

Annotation: Based on Steiner's approach, this book stresses the importance of the "life process (biological process)," with the farm or garden a biological organic unit, not a series of unconnected processes. JPG

Cited in: Coleman (1976); Conford (2001); Harwood (1983); Harwood (1990); Merrill (1983)

1938

United States Department of Agriculture

Soils and Men: The Yearbook of Agriculture, 1938

Washington DC: U.S. Government Printing Office, 1938. 1232p. Includes index. Glossary. Literature cited.

NAL Call no: 1 AG844 1938

Annotation: More than 100 authors contributed to this yearbook, including William A. Albrecht, then Professor of Soils at the University of Missouri and one of the fathers of the ecological agricultural movement. It represents an effort to see "the subject as a whole—scientific aspects, practical aspects, social and economic aspects; the needs of individuals, groups and the Nation." JPG

Cited in: Blum (1993); Conford (2001); Merrill (1983)

1939

Jacks, Graham Vernon, 1901-1977; and Robert Orr Whyte, 1903-1986

The Rape of the Earth: A World Survey of Soil Erosion

London: Faber and Faber, 1939. 313p. Includes index. 47 illustrations. Published in America under the title *Vanishing Lands; a World Survey of Soil Erosion.*

NAL Call no: 56.7 J13

Other works by Jacks: *Soil* (1954). Other works by Whyte: *Crop Production and Environment* (1946; revised, 1960); *Land, Livestock and Human Nutrition in India* (1968); *Tropical Grazing Lands: Communities and Constituent Species* (1974).

Annotation: Frequently cited, this book is a pioneering classic on the subject. JPG

Cited in: Balfour (1943); Conford (1988); Conford (2001); Howard (1940); Merrill (1983); Northbourne (1940); Scofield (1986)

1939

Price, Weston Andrew, 1870-1948

Nutrition and Physical Degeneration; a Comparison of Primitive and Modern Diets and Their Effects

New York; London: P. B. Hoeber, 1940 (oldest edition held by the National Agricultural Library). xviii, 431p. Foreword by Earnest Albert Hooton. References at end of some of the chapters. Includes 134 figures (primarily photographs taken by the author). Other editions: A 50th anniversary edition, 1989, includes forewords from the original editions by Earnest Albert Hooton, Granville Frank Knight, M.D., William A. Albrecht, Ph.D., and new introductions and reminiscences specially written for the Golden Anniversary Edition by Abram Hoffer, M.D., Ph.D., H. Leon Abrams, Jr. and Donald Delmage Fawcett.

NAL Call no: 389.1 P93

Full-text: Journey to Forever Online Library, http://journeytoforever.org/farm_library.html (accessed Jan. 1, 2007)

Annotation: A practicing dentist, Price set out to discover why certain "primitive" peoples exhibited perfect teeth while the majority of individuals from modern societies had such poor ones. His travels and work of the 1920s and 1930s produced broad nutritional studies that linked many of modern society's health problems to diet and to how modern food is grown and prepared. MVG

Cited in: Conford (2001); Merrill (1983)

1940

Bruce, Maye Emily, 1879-

From Vegetable Waste to Fertile Soil (Quick Return Compost)

London: C.A. Pearson, 1940. 64p. Bibliography, p. 63. Foreword by L.F. Easterbrook.

NAL Call no: 57.4 B83

Other works by this author: *Common-sense Compost Making by the Quick Return Method* (1946) (full-text: Journey to Forever Online Library, http://journey-toforever.org/farm_library.html) (accessed Apr. 23, 2007); Compost Making: Practical Advice on Nature's Method of Restoring Life in the Soil (1947).

Annotation: Bruce was a composting pioneer, along with Sir Albert Howard. Both became founding members of the Soil Association in England. MVG

Cited in: Coleman (1976); Conford (2001); Harwood (1983); Harwood (1990)

1940

Howard, Sir Albert, 1873-1947

An Agricultural Testament

Oxford: Oxford University Press, 1940. 253p. Includes index. Literature. Bibliographies at ends of chapters.

NAL Call no: 56.6 H83A

Full-text: Soil and Health Library, Steve Solomon; and Journey to Forever Online Library, http://www.soilandhealth.org/01aglibrary/01principles.html; http://journeytoforever.org/farm_library.html (accessed Jan. 1, 2007)

Other works by this author: *Soil and Health: A Study of Organic Agriculture* (1947).

Annotation: This is the classic study on soil fertility by the "father of the movement." It includes the "Agriculture of the Nations Which Have Passed Away" and observations of agricultural practices of both the Orient and the Occident. JPG

Cited in: Balfour (1943); Blum (1993); Coleman (1976); Conford (1988); Conford (2001); Esbjornson (1992); Harwood (1983); Harwood (1990); Heckman (2006); Kirschenmann (2004); Korcak (1992); Kuepper, Gegner (2004); Merrill (1983); Northbourne (1940); Scofield (1986)

1940

Northbourne, Walter Ernest Christopher James Lord Baron, 1896-1982

Look to the Land

London: Dent, 1940. 206p. Includes index. Bibliography.

NAL Call no: 30 N81

Other works by this author: *Religion in the Modern World* (1963); *Looking Back on Progress* (1970).

Annotation: This frequently overlooked early inspirational work includes the first known use of the term "organic farming" in a chapter heading on page 148, "diversified organic farming a practical proposition." According to his son, the present Baron, Northbourne felt obliged, when World War II came in 1939, "to recommend to other farms the chemical methods of stimulating production" in order to "help feed" the country. "Being an honourable person, he therefore felt that he, too, must abandon his organic production and adopt the more conventional methods of fertilising and weed control which were beginning to emerge at that time." (Personal correspondence 9/88) JPG

Cited in: Blum (1993); Coleman (1976); Conford (2001); Harwood (1983); Harwood (1990); Kirschenmann (2004); Korcak (1992); Scofield (1986)

1941

Beeson, Kenneth Crees, 1903-1998

The Mineral Composition of Crops with Particular Reference to the Soils in Which They Were Grown: A Review and Compilation

Washington DC: United States Department of Agriculture, 1941. 164p. "The work represented by this publication was supported by the Bankhead-Jones Research Fund. It was initiated in the Bureau of Chemistry and Soils (now the Bureau of Agricultural Chemistry and Engineering) and was later transferred to the Bureau of Plant Industry." Bibliography, p. 59-90. "Sources of unpublished material," p. 91. (Miscellaneous Publication, 369)

NAL Call no: 1 Ag84M no.369

Full-text: Organic Roots, Organic Agriculture Information Access, http://www.hti.umich.edu/n/nal/ (to be added June, 2007) (accessed Jan. 1, 2007)

Annotation: Although he found "confusion and contradictory results" in the fertilizer studies he read, Beeson did arrive at a couple of important conclusions with this review. One, "...empirical investigations have quite definitely shown that Liebig's 'law of the minimum' never represented the mechanism of absorption of nutrients by plants and that the actual facts seem to indicate that when one of the principal nutrients is deficient in the soil solution, the others are taken up by the plant in amounts greater than normal..." ; and two, that "...fundamental studies of what changes take place in the soil when the fertilizers are applied and the effect these changes will have on the plant are lacking." Scientists continue to struggle with researching these same issues. MVG

1941

Graham, Michael
Soil and Sense
London: Faber and Faber, 1941. 274p. Preface by Sir E. John Russell. "This book grew out of articles in Riding [magazine]."
NAL Call no: 32 G762
Full-text: Soil and Health Library, Steve Solomon, http://www.soilandhealth.org/01aglibrary/01principles.html (accessed Jan. 1, 2007)
Other works by this author: *A Natural Ecology* (1973).
Annotation: In Great Britain, the unprecedented cultivation of traditional pastureland and the introduction of "modern" tillage and fertilization practices during World War II provided much needed food. However, Graham and many others saw this as "land-robbery," and feared for the future of soils and food production there. Graham emphasized the need to return to integrated grassland and animal production and stated, "...whenever the public purse is used to help farming, that the money should go to good farmers who feed the land as tradition says it should be fed and that none should go to those who merely exploit it." MVG
Cited in: Coleman (1976); Harwood (1983); Harwood (1990)

1942

Barlow, Kenneth Elliott, 1906-2000
The Discipline of Peace
London: C. Knight, 1971 (oldest edition held by the National Agricultural Library). 147p. Other editions: 2nd edition, 1971, includes a new Introduction by the author and preface by Robert Waller. (Classics of Human Ecology, 2)
NAL Call no: GN320.B3 1971

Other works by this author: *A Home of Their Own* (1946); *The Law and the Loaf* (1978); *Soil, Food and Health in a Changing World*, co-edited with Peter Bunyard (1981); *Recognising Health* (1988).

Annotation: A medical doctor in Britain, Barlow completed this publication during World War II. It examines the ecological basis of a sustainable and "fulfilling civilization" in the context of post-war reconstruction. "Current concerns about food quality and the dangers of a technological attitude to the environment were anticipated 60 years ago by Barlow and a farsighted group of agriculturalists, doctors and writers. They developed an organic philosophy and established the Soil Association to put principles into practice; Barlow was a founder member." "Obituary: Dr Kenneth Barlow," by Philip Conford, in The Independent (London), Mar 1, 2001. http://www.findarticles.com/p/articles/mi_qn4158/is_20010301/ai_n14375583 (accessed Feb. 1, 2007) MVG

Cited in: Conford (2001); Harwood (1983); Harwood (1990)

1942

Billington, Francis Howard, -1947

Compost for Garden Plot or Thousand-acre Farm: A Practical Guide to Modern Methods

London: Faber and Faber, 1942. 88p. Includes a glossary and annotated "Brief Bibliography of Non-technical Works," p. 85-88. Other editions: 4th edition of this work, revised by Ben Easey, 1956.

NAL Call no: 57.4 B49

Annotation: Written in Britain during World War II, during a shortage of chemical fertilizers, this small volume organized a growing body of information about composting—bio-dynamic practices in Germany, Sir Albert Howard's Indore methods and the "Quick Return" compost system—into an instructive manual for gardeners and farmers. MVG

Cited in: Coleman (1976); Harwood (1983); Harwood (1990)

1943

Balfour, Lady Evelyn Barbara, 1899-1990

The Living Soil

London: Faber and Faber, 1943. 248p. Includes index. Glossary. Bibliography. Other editions: The Living Soil; Evidence of the Importance to Human Health of Soil Vitality, with Special Reference to Post-war Planning (1944); The Living Soil and the Haughley Experiment (1975).

NAL Call no: 56.5 B19

Full-text: Soil and Health Library, Steve Solomon; and Journey to Forever Online Library, http://www.soilandhealth.org/01aglibrary/01principles.html; http://journeytoforever.org/farm_library.html (accessed Jan. 1, 2007)

Other works by this author: *Towards a Sustainable Agriculture—The Living Soil* (speech presented at the 1977 International Federation of Organic Agriculture Movements (IFOAM) conference in Switzerland) (full-text: Journey to Forever Library, http://journeytoforever.org/farm_library/balfour_sustag.html) (accessed Apr. 23, 2007).

Annotation: Based on 32 years' comparison of organic, mixed and chemical sections of a farm at Haughley, England, this is an extremely readable exposition of the evidence in favor of biological agriculture by one of the founders of that country's "Soil Association." JPG

Cited in: Blum (1993); Coleman (1976); Conford (1988); Conford (2001); Harwood (1983); Harwood (1990); Kirschenmann (2004); Korcak (1992); Merrill (1983); Scofield (1986)

1943

Faulkner, Edward Hubert, 1886-1964

Plowman's Folly

New York: Grossett and Dunlap, 1943. 155p.

NAL Call no: 56.7 F27

Full-text: Soil and Health Library, Steve Solomon; and Journey to Forever Online Library, http://www.soilandhealth.org/01aglibrary/01principles.html; http://journeytoforever.org/farm_library.html (accessed Jan. 1, 2007)

Other works by this author: *Practical Farming for the South,* with B. F. Bullock (1944); *Uneasy Money* (1946); *A Second Look* (1947); *Ploughing in Prejudices* (1948); *Soil Restoration* (1953).

Annotation: As the title implies, this famous work states that plowing is wrong and that the moldboard plow is not a satisfactory implement for the preparation of land for the production of crops. JPG

Cited in: Beeman (1993); Coleman (1976); Conford (2001); Harwood (1983); Harwood (1990); Heckman (2006)

1944

Henderson, George

The Farming Ladder

London; Boston: Faber and Faber, 1974 (oldest edition held by the National Agricultural Library). 246p.

NAL Call no: S455.H423-1978

Full-text: Soil and Health Library, Steve Solomon, http://www.soilandhealth.org/01aglibrary/01principles.html (accessed Jan. 1, 2007)

Other works by this author: *Farmer's Progress* (1950) (full-text: Soil and Health Library, Steve Solomon, http://www.soilandhealth.org/01aglibrary/01principles.html) (accessed Apr. 23, 2007); *The Farming Manual* (1960).

Annotation: Henderson's readable books describe farming from the farmer's point of view. "The preservation of fertility is the first duty of all that live by the land... There is only one rule of good husbandry—leave the land far better than you found it." Henderson's books are highly recommended by Eliot Coleman. MVG

Cited in: Coleman (1976)

1945–1979

TWO SEPARATE PATHS

Author Joseph Heckman notes that the "period from about 1940 to 1978 may be called the era of polarization of agriculture into organic and non-organic camps." ("A History of Organic Farming—Transitions from Sir Albert Howard's War in the Soil to the USDA National Organic Program." Renewable Agriculture and Food Systems (2006), vol, 21, no. 3, pp. 143-150 (full-text at Weston A. Price Foundation, http://www.westonaprice.org/farming/history-organic-farming.html) (accessed May 7, 2007)

Mainstream agriculture emphasized improving crop and livestock production volume at every level. Dramatic increases in yield per acre were made and the "Green Revolution" with its breeding of high-yielding grain varieties, monocropping, and extensive use of energy, fertilizer and pesticides came to symbolize farming progress. The agricultural research establishment distanced itself from organic approaches and those who espoused them.

Nonetheless, a small contingent of scientists, farmers, writers and others continued to explore ecological soil building techniques, gardening without synthetic chemicals and other "alternatives" to conventional farming and marketing. The environmental movement's onset in the late 1960s and early 1970s marked the beginning of the end of this polarization.

1945

Delbet, Pierre Louis Ernest, 1861-1957

L'Agriculture et la Santé

Paris: Denoel, 1945. 111p. Translated title: Agriculture and Health.

NAL Call no: 57 D372

Annotation: Dr. Delbet was a physician who experimented at length with the medicinal applications of magnesium chloride. His studies led him explore the relationship between human health and agriculture, and he espoused "biological" food production systems. He is credited with the first use of the French term, "agriculture biologique." ("biological agriculture"). In the 1950s, Raoul Lemaire and Jean Boucher built on Delbet's concept of biological agriculture, creating the widely adopted Lemaire-Boucher system. MVG. See also: Boucher, 1968.

1945

Rayner, Mabel Cheveley

Trees and Toadstools

London: Faber and Faber, 1946 (oldest edition held by the National Agricultural Library). 71p. First published in 1945. Republished, 1947.

NAL Call no: 463.88 R21T

Full-text: Soil and Health Library, Steve Solomon; and Journey to Forever Online Library, http://www.soilandhealth.org/01aglibrary/01principles.html; http://journeytoforever.org/farm_library.html (accessed Jan. 1, 2007)

Other works by this author: *Problems in Tree Nutrition* (1944).

Annotation: This non-technical book discusses early mycorrhizal research. MVG

Cited in: Merrill (1983)

1945

Rodale, Jerome Irving, 1898-1971

Pay Dirt: Farming and Gardening with Composts

New York: Devin Adair, 1945. 242p. Bibliography. Introduction by Albert Howard.

NAL Call no: 57.4 R61

Full-text: Core Historical Literature of Agriculture, Cornell University, http://chla.library.cornell.edu/cgi/t/text/text-idx?c=chla;idno=2838381 (accessed Jan. 1, 2007)

Other works by this author: *Organic Front* (1948); *Stone Mulching in the Garden* (1949); *Organic Method on the Farm* (1949); *Organic Gardening: How to Grow Healthy Vegetables, Fruits, and Flowers Using Nature's Own Methods* (1955); *Garden Success without Poison Sprays* (1962); *The Prevention System for Better Health,* with Robert Rodale (1974).

Annotation: The classic statement on the value of soil, this and Rodale's later works sparked and fueled the organic movement in North America. JPG. See also: Rodale, 1948.

Cited in: Beeman (1993); Conford (2001); Harwood (1983); Harwood (1990); Heckman (2006); Korcak (1992); Kuepper, Gegner (2004); Rateaver (1973), Scofield (1986)

1945

Shepard, Ward, 1887-

Food or Famine, the Challenge of Erosion

New York: Macmillan, 1945. x, 225p. Includes index.

NAL Call no: 56.7 Sh4F

Full-text: Core Historical Literature of Agriculture, Cornell University, http://chla.library.cornell.edu/cgi/t/text/text-idx?c=chla;idno=3135913 (accessed Jan. 1, 2007)

Other works by this author: *Forests and Floods* (1928); *The Forest Problem* (1929); *Outline of a Proposed Organic Act of Congress to Prevent Forest Degeneration and Destruction and to Preserve and Rebuild Forest Resources* (1940); *La Conservacion de las Tierras Indigenas en los Estados Unidos* (for the National Indian Institute, Department of the Interior, 1942).

Annotation: In the wake of the Dust Bowl, Shepard wrote eloquently about soil conservation. "Erosion, floods, droughts, the destruction of great land masses, the violent disruption of whole river systems—these are nature's weapons against the impious creature who dares rob her riches without replenishing them." His work pointed to the relationship of soil loss to water and irrigation practices, farming and forestry, land ownership patterns and more. "World agriculture must be established on biological and ecological principles that emulate nature's way of maintaining the dynamic soil-water-plant complex." MVG

Cited in: Beeman (1993)

1946

Kolisko, Eugen, 1893-1939; and Lily Noha Kolisko, 1889-

Agriculture of Tomorrow

Gloucester: Kolisko Archive, 1946. 426p. Includes index. Bibliography. Other editions: 2nd edition, 1978.

NAL Call no: 30 K833

Full-text: (1939 edition) Soil and Health Library, Steve Solomon, http://www.soilandhealth.org/01aglibrary/01principles.html (accessed Jan. 1, 2007)

Annotation: This account of scientific work based on the principles of Rudolf Steiner details the Koliskos' controlled experiments on the effects of cosmic forces on crop production. JPG

Cited in: Conford (2001); Harwood (1983)

1946

Wrench, Guy Theodore, -1954

Reconstruction by Way of the Soil

London: Faber and Faber, 1946. 262p.

NAL Call no: 56 W92

Full-text: Journey to Forever Online Library, http://journeytoforever.org/farm_library.html (accessed Jan. 1, 2007)

Other works by this author: *The Wheel of Life: A Study of Very Healthy People* (1938) (full-text: Journey to Forever Online Library, http://journeytoforever.org/farm_library.html) (accessed Jan. 1, 2007).

Annotation: This book presents an international history of how the earth's soil resources have been used and abused through the centuries, from Roman times through World War II. Wrench calls on post-war society to wage "A new social, non-military war... war on behalf of the soil and of the healthy life and physical freedom of men." (Chapter 24, "Action") Wrench also authored an influential book on human nutrition, The Wheel of Life: A Study of Very Healthy People (C.W. Daniel, 1938), a study of the Hunza, a mountain people renowned for their longevity and vigor. Wrench postulated that human health depends on a "whole" diet that emphasizes consumption of natural foods grown in an environmentally sound way. MVG

Cited in: Coleman (1976); Conford (1988); Harwood (1983)

1947

Bromfield, Louis, 1896-1956

Malabar Farm

New York: Harper, 1947. 405p. Drawings by Kate Lord.

NAL Call no: 31.3 B78M

Other works by this author: *Pleasant Valley* (1945); *Few Brass Tacks* (1946); *Out of the Earth* (1950); *New Pattern for a Tired World* (1954); *From my Experience; the Pleasures and Miseries of Life on a Farm* (1955); *Animals and Other People* (1955); *Wealth of the Soil* (1959).

Annotation: The journal of the famous Ohio farmer/author, this work intersperses anecdotes and history with practical advice and a recounting of his experiences. It is a sequel to his earlier book, Pleasant Valley, published in 1945. JPG

Cited in: Beeman (1993); Blum (1993); Coleman (1976); Conford (2001); Harwood (1983), Harwood (1990); Heckman (2006); Merrill (1983); Rateaver (1973)

1947

Howard, Lady Louise Ernestine Matthaei, 1880-

The Earth's Green Carpet

London: Faber and Faber, 1947. 219p. Includes index. Appendices, including "The Indore Process and its Evolution" and a "List of Books" (including periodicals).

NAL Call no: 30 H83

Full-text: Soil and Health Library, Steve Solomon; and Journey to Forever Online Library, http://www.soilandhealth.org/01aglibrary/01principles.html; http://journeytoforever.org/farm_library.html (accessed Jan. 1, 2007)
Other works by this author: *Labour in Agriculture: An International Survey* (1935); *Sir Albert Howard in India* (1953) (full-text: Soil and Health Library, Steve Solomon, http://www.soilandhealth.org/01aglibrary/01principles.html) (accessed Apr. 23, 2007).
Annotation: This is an account of the ideas and principles of Sir Albert Howard written by his second wife, also a proponent of organic agriculture. JPG
Cited in: Coleman (1976); Harwood (1983); Merrill (1983)

1948

Lowdermilk, Walter Clay, 1888-1974
Conquest of the Land Through Seven Thousand Centuries
Washington: Soil Conservation Service, U.S. Dept. of Agriculture, 1975 (oldest edition held by the National Agricultural Library). 30p. Includes illustrations and photographs taken by Dr. Lowdermilk. (Agriculture Information Bulletin, 99)
NAL Call no: 1 Ag84Ab no.99 1975
Full-text: National Agricultural Library Digital Repository (NALDR) (1953 version), http://naldr.nal.usda.gov/NALWeb/Search.aspx (search on "lowdermilk") (accessed Jan. 1, 2007)
Other works by this author: *Erosion Control in Japan* (1934); *History of Soil Use in the Wu t'ai Shan Area,* with Dean R. Wickes (1938); *Soil, Forest, and Water Conservation and Reclamation in China, Israel, Africa, and the United States; an Interview Conducted by Malca Chall* (1969).
Annotation: "In 1938 and 1939, Dr. Lowdermilk, formerly Assistant Chief of Soil Conservation Service, made an 18-month tour of western Europe, North Africa and the Middle East to study soil erosion and land use in those areas... The main objective of the tour was to gain information from those areas - where some lands had been in cultivation for hundreds and thousands of years - that might be of value in helping to solve the soil erosion and land use problems of the United States." Preface. MVG
Cited in: Blum (1993); Merrill (1983); Worster (1985)

1948

Okada, Mokichi, 1882-1955

Fertilizer-free Agriculture

Publisher unknown: 1948. 96p.

NAL Call no: n.a.

Other works by this author: *Health and the New Civilization* (1991).

Annotation: A Japanese spiritual leader, Okada believed that medical science and agriculture play the most vital roles in the maintenance of life and health. Although English-language documentation for early publications is lacking at the National Agricultural Library, Okada's farming system dates from 1936; his original work (ca. 1946) was called, A Great Agricultural Revolution. Okada's philosophy is now associated with "Kyusei Nature Farming" and "Shumei Natural Agriculture." MVG

1948

Osborn, Fairfield, 1887-1969

Our Plundered Planet

Boston: Little, Brown, 1948. 217p. Includes bibliography.

NAL Call no: 279 Os1

Full-text: Core Historical Literature of Agriculture, Cornell University, http://chla.library.cornell.edu/cgi/t/text/text-idx?c=chla;idno=2932687 (accessed Jan. 1, 2007)

Other works by this author: *60,000 More Every 24 Hours!* (1951); *The Limits of the Earth* (1953).

Annotation: Dedicated "to all who care about tomorrow," Osborn's book places the post-World War II environmental situation in historical and political context, from citing the problems of bygone civilizations to speculating about the future of Soviet farming. It is "an attempt to show what man has done in recent centuries to the face of the earth and the accumulated velocity with which he is destroying his own life sources." MVG

Cited in: Beeman (1993)

1948

Rodale, Jerome Irving, 1898-1971

The Organic Front
Emmaus PA: Rodale Press, 1948. 198p.
NAL Call no: 56.6 R610
Full-text: Soil and Health Library, Steve Solomon, http://www.soilandhealth.org/01aglibrary/01principles.html (accessed Jan. 1, 2007)
Other works by this author: *Pay Dirt: Farming and Gardening with Composts* (1945); *Stone Mulching in the Garden* (1949); *Organic Method on the Farm* (1949); *Organic Gardening: How to Grow Healthy Vegetables, Fruits, and Flowers Using Nature's Own Methods* (1955); *Garden Success without Poison Sprays* (1962); *The Prevention System for Better Health,* with Robert Rodale (1974).
Annotation: J. I. Rodale called the movement "organiculture," the "new, yet age-old method." JPG. See also: Rodale, 1945.
Cited in: Beeman (1993); Conford (2001); Harwood (1983); Harwood (1990); Heckman (2006); Merrill (1983)

1949
Blackburn, John Stead, compiler, 1882-
Organic Husbandry: A Symposium
London: The Biotechnic Press, 1949. 160p. Annotated bibliography, p. 149-160.
NAL Call no: 56.6 B56
Annotation: This unique compendium of papers includes such diverse sources as Sir Albert Howard's essay, "The Fresh Produce of Fertile Soil is the Real Basis of Public Health," "My Compost Garden," by L.F. Easterbrook and "Is Digging Necessary?" by F. C. King. The appended bibliography of historical and contemporary books is delightfully annotated by the compiler. MVG
Cited in: Conford (2001)

1949
Butler, Lowell F.
The Decreasing Fertility of Western Soil
Denver: Atlas Printing and Engraving Co., 1949. 39p. Foreword by A. R. Bunger.
NAL Call no: 56.6 B97
Annotation: Butler, an agricultural educator and plant pathologist, addressed, "soil exhaustion" and its remedies specific to calcareous, irrigated soils in the western United States. MVG

1949

Leopold, Aldo Carl, 1886-1948

A Sand County Almanac and Sketches Here and There

New York: Oxford University Press, 1949. 226p. Illustrated by Charles W. Schwartz.

NAL Call no: 409 L552

Other works by this author: *Game Management* (1933); *Ecological Conscience* (1947); *Round River; From the Journals of Aldo Leopold,* edited by Luna B. Leopold (1953); *The River of the Mother of God and Other Essays,* edited by Susan L. Flader and J. Baird Callicott (1991); *For the Health of the Land: Previously Unpublished Essays and Other Writings,* edited by J. Baird Callicott and Eric T. Freyfogle (1999).

Annotation: In this well-known collection of essays, Leopold presents the case for a land ethic as a product of social evolution. JPG

Cited in: Beeman (1993); Esbjornson (1992); Lehman (1993)

1949

Picton, Lionel James, 1874-

Nutrition and the Soil, Thoughts on Feeding

New York: Devin Adair, 1949. ix, 374p. Introductory essay on creative medicine, by Jonathan Forman. First published in Great Britain as Thoughts on Feeding.

NAL Call no: 56.6 P58

Full-text: Core Historical Literature of Agriculture, Cornell University, http://chla.library.cornell.edu/cgi/t/text/text-idx?c=chla;idno=3134147 (accessed Jan. 1, 2007)

Annotation: Picton, a medical doctor, was the person who "linked [Sir Albert] Howard's work on crop breeding and Sir Robert McCarrison's studies in nutrition. He brought the two men together in 1939 for the launch of the 'Medical Testament,' a document urging investigation of the relationship between compost-grown food and human health." "Organic Origins: The Ideas Shaping the Soil Association 60 Years Ago," by Phillip Conford, in Living Earth Newsletter, Spring 2006 (http://www.soilassociation.org/web/sa/saweb.nsf/848d689047cb466780256a6b00298980/bb02cc7c62b6ce7c802571a30042f3ef//Ideas%20shaping%20the%20Soil%20Association%2060%20years%20ago.pdf) (accessed Feb. 1, 2007). MVG

Cited in: Coleman (1976); Rateaver (1973)

1950

Cocannouer, Joseph A.

Weeds: Guardians of the Soil

New York: Devin Adair, 1950. 179p. Brief list of "other books on organiculture" opposite title page.

NAL Call no: 79 C04

Full-text: Journey to Forever Online Library, http://journeytoforever.org/farm_library.html (accessed Jan. 1, 2007)

Other works by this author: *Farming with Nature* (1954; reissued as *Organic Gardening and Farming*, 1997); *Water and the Cycle of Life* (1958).

Annotation: Called a "pioneering work" in the advocacy of the controlled use of weeds, this book's author wrote on other subjects having to do with modern farming practices. JPG

Cited in: Coleman (1976); Harwood (1983); Harwood (1990); Merrill (1983); Rateaver (1973)

1950

University of California Berkeley Sanitary Engineering Research Laboratory

Composting for Disposal of Organic Refuse

Berkeley CA: Sanitary Engineering Research Laboratory, 1950. 42p. (Technical Bulletin, 1)

NAL Call no: 57.4 Cl2

Annotation: This study examines a Berkley CA modification of the Becari municipal composting system, successfully instituted in Italy. Although U.S. interest in commercial city composting systems peaked and then subsided in the early 1950s, the legacy of European and U.S. experiments from this period served a successful revival of large-scale composting efforts instituted in the 1980s and 1990s. MVG

Cited in: Blum (1993)

1951

Sykes, Friend Frank, 1888-1965

Food, Farming and the Future

London: Faber and Faber, 1951. Introduction by Louise E. Howard.

NAL Call no: 32 Sy4F

Other works by this author: *This Farming Business* (1944); *Humus and the Farmer* (1946) (full-text: Soil and Health Library, Steve Solomon, http://www.soilandhealth.org/01aglibrary/01principles.html) (accessed Apr. 23, 2007); *Living from the Land: A Guide to Farm Management* (1957); *Modern Humus Farming* (1959).

Annotation: An experienced British farmer, the author of this book also wrote Humus and the Farmer (1946) and Modern Humus Farming (1959). JPG

Cited in: Coleman (1976); Conford (1988); Conford (2001); Harwood (1983); Harwood (1990); Korcak (1992); Merrill (1983)

1952

Hyams, Edward Solomon, 1912 -1975

Soil and Civilization

London; New York: Thames and Hudson, 1952. 312p. Includes bibliography. Republished, Harper and Row, 1976. (The Past in the Present Series)

NAL Call no: 56 H99

Other works by this author: *Plants in the Service of Man 10,000 Years of Domestication* (1971).

Annotation: A classic book that describes the parallel relationship between the decline of many of the world's great civilizations and the degradation of their soil resources. MVG

Cited in: Conford (1988); Conford (2001); Merrill (1983); Worster (1985)

1953

Odum, Eugene Pleasants, 1913-2002; and Howard Thomas Odum, 1924-2002

Fundamentals of Ecology

Philadelphia: Saunders, 1953. 384p. Includes bibliography. Other editions: 2nd, 1959; 3rd, 1971; 5th, with Gary Barrett, 2004.

NAL Call no: 442 Od8

Other works by Eugene Odum: *Ecology and Our Endangered Life-support Systems* (1989). Other works by Howard Odum: *Environment, Power, and Society* (1971);

Energy Basis for Man and Nature, with Elizabeth C. Odum (1976); *Ecological and General Systems: An Introduction to Systems Ecology* (1994); *Environmental Accounting: EMERGY and Environmental Decision Making* (1996).

Annotation: This book was one of the first textbooks to embrace a holistic view of ecosystems and human activities. Systems ecology pioneer, Eugene Odum, is listed as the sole author on the first edition of this work; collaboration with his brother Howard commenced with succeeding editions. Over the years, both brothers educated a wide audience about ecological and environmental issues. MVG

1954

Hainsworth, Peter Hugh, 1921-

Agriculture, a New Approach

London: Faber and Faber, 1954. 248p. Includes index. Bibliography. Glossary. Republished as Agriculture, the Only Right Approach: Science Says There is a Difference (Rateaver, 1976).

NAL Call no: 30 H12

Annotation: One of the best early sources of technical information regarding natural fertilizing, this represents an "attempt to gather together relevant material that may have some bearing on the results achieved by,..organic methods." JPG

Cited in: Blum (1993); Coleman (1976); Harwood (1983); Harwood (1990); Kuepper, Gegner (2004); Merrill (1983)

1954

Nearing, Helen, 1904-1995; and Scott Nearing, 1883-1983

Living the Good Life: Being a Plain Practical Account of a Twenty Year Project in a Self-subsistent Homestead in Vermont, Together with Remarks on How to Live Sanely and Simply in a Troubled World

Harborside ME: Social Science Institute, 1954. 209p. Includes bibliography. Reprinted, 1970.

NAL Call no: 281.2 N27

Other works by these authors: *The Maple Sugar Book, Being a Plain, Practical Account of the Art of Sugaring Designed to Promote an Acquaintance with the*

Ancient as Well as the Modern Practice, Together with Remarks on Pioneering as a Way of Living in the Twentieth Century (1950); *Building and Using our Sun-heated Greenhouse: Grow Vegetables All Year-round* (1977); *Continuing the Good Life* (1979).

Annotation: The Nearings became mentors to a generation of back-to-the-landers and modern homesteaders during the late 1960s and 1970s. The couple's inspirational writings were based on practical experience gained from life on their organic farmsteads in Vermont and Maine. MVG

Cited in: Rateaver (1973)

1954

Wickenden, Leonard

Gardening with Nature: How to Grow Your own Vegetables, Fruits and Flowers by Natural Methods

New York: Devin Adair, 1954. 392p. Includes index. Suggested readings.

NAL Call no: 90 W632

Other works by this author: *Make Friends with Your Land: A Chemist Looks at Organiculture* (1949); *Our Daily Poison; the Effects of DDT, Fluorides, Hormones and Other Chemicals on Modern Man* (1955).

Annotation: Depicted by the author, who has a chemistry background, as largely a "how-to" book, this work integrates Wickenden's advice with the scientific basis on which the organic concept of plant growth was built. Readers can thus learn both organic gardening practices and how to defend them. In 1949 Wickenden had written the popular Make Friends with Your Land: A Chemist Looks at Organiculture. JPG

Cited in: Coleman (1976); Harwood (1983); Harwood (1990); Rateaver (1973)

1954

Yeomans, Percival Alfred, 1905-1984

The Keyline Plan

Sydney: P.A. Yeomans, 1954. 120p. Also entitled The Australian Keyline Plan.

NAL Call no: 282 1992 Y4K

Full-text: Soil and Health Library, Steve Solomon, http://www.soilandhealth.org/01aglibrary/01principles.html (accessed Jan. 1, 2007)

Other works by this author: *The City Forest: The Keyline Plan for the Human Environment Revolution* (1951); *Challenge of Landscape: the Development and Practice of Keyline* (1958) (both in full-text: Soil and Health Library, Steve Solomon, http://www.soilandhealth.org/01aglibrary/01principles.html) (accessed Apr. 23, 2007); *Water for Every Farm* (1965, updated and reissued by Ken Yeomans in 1993).

Annotation: P.A. Yeomans pioneered "the use of on farm irrigation dams in Australia, as well as chisel plows and subsoil aerating rippers. Yeomans perfected a system of amplified contour ripping that controlled rainfall run off and enabled the fast flood irrigation of undulating land with out the need for terracing." Keyline Designs Web site (http://www.keyline.com.au/) (accessed Apr. 23, 2007). MVG

1955

Easey, Ben, 1925-

Practical Organic Gardening

London: Faber and Faber, 1955. 151p. Includes index. Bibliography. Appendices. Lists of gardens, suppliers, and organizations. Other editions: Revised, 1976.

NAL Call no: 57.4 Ea7

Annotation: This is a substantive work on the subject, with a thorough and well-documented framework. JPG

Cited in: Coleman (1976); Conford (2001)

1955

Stout, Ruth, 1891-1980

How to Have a Green Thumb without an Aching Back: A New Method of Mulch Gardening

New York: Exposition Press, 1955. 164p. Reissued in 1987.

NAL Call no: SB453.S76 1955 DNAr

Other works by this author: *Gardening without Work; for the Aging, the Busy, and the Indolent* (1961, republished 1998); *The Ruth Stout No-work Garden Book* (1971).

Annotation: Although mulch gardening was not really "new" in the 1950s, this popular work defined and encouraged no-till techniques for small-scale growers. It renewed interest in soil biological processes and organic gardening for a new generation. MVG
Cited in: Rateaver (1973)

1955

Turner, Newman, 1913-

Fertility Pastures: Herbal Leys as the Basis of Soil Fertility and Animal Husbandry

London: Faber and Faber, 1955. 204p. Includes index.

NAL Call no: 60.1 T85

Full-text: Soil and Health Library, Steve Solomon, http://www.soilandhealth.org/01aglibrary/01principles.html (accessed Jan. 1, 2007)

Other works by this author: *Fertility Farming* (1951); *Herdsmanship* (1952) (both in full-text: Soil and Health Library, Steve Solomon, http://www.soilandhealth.org/01aglibrary/01principles.html) (accessed Apr. 23, 2007).

Annotation: Also spelled "lea"; the dictionary defines ley as "arable land sown to grasses or clover for hay or grazing and usually plowed and planted with other crops after two or three years." Turner boasts that the "herbal ley is my manure merchant, my food manufacturer and my vet, all in one." JPG

Cited in: Blum (1993); Coleman (1976); Conford (2001); Harwood (1983); Merrill (1983)

1957

Gilbert, Frank Albert, 1900-

Mineral Nutrition and the Balance of Life

Norman OK: University of Oklahoma Press, 1957. xv, 350p. Bibliography, p. 264-336.

NAL Call no: 386.3 G37

Other works by this author: *Mineral Nutrition of Plants and Animals* (1948) (full-text: Core Historical Literature of Agriculture, http://chla.library.cornell.edu/cgi/t/text/text-idx?c=chla;idno=2837840) (accessed Feb. 1, 2007); Metal Trace Elements in Agriculture (1949).

Annotation: This encyclopedic review of research and literature concerning mineral nutrition was the first compilation if its kind. Its breadth, extensive bibliography and illustrative photographs continue to make it a valuable tool in understanding soil-plant, animal and human nutrition relationships. MVG

Cited in: Blum (1993); Merrill (1983)

1957

Rowe-Dutton, Patricia

The Mulching of Vegetables

Farnham Royal, England: C.A.B. International Bureau of Horticulture and Plantation Crops, 1957. 169p. (Technical Communication, 24)

NAL Call no: 84 IM72

Annotation: This is a comprehensive review of research about mulching including mulching's relationship to plant diseases and erosion in vegetable production. MVG

Cited in: Coleman (1976)

1957

Voisin, André, 1903-1964

Grass Productivity

New York: Philosophical Library, 1959. 353p. Translated from the French, Productivité de l'Herbe, by Catherine T. M. Herriot. Includes bibliography. Reprinted, 1988.

NAL Call no: 60.1 V87Ge

Other works by this author: *Soil, Grass, and Cancer: Health of Animals and Men is Linked to the Mineral Balance of the Soil* (1959, republished by Acres U.S.A., 1999); *Better Grassland Sward; Ecology, Botany, Management* (1960); *Rational Grazing: The Meeting of Cow and Grass; a Manual of Grass Productivity,* with Antoine Lecomte (1962); *Grass Tetany* (1963?) (full-text: Soil and Health Library, Steve Solomon, http://www.soilandhealth.org/01aglibrary/01principles.html) (accessed Apr. 23, 2007); *Fertilizer Application: Soil, Plant, Animal* (Three of 25 lectures, entitled Soil, Plant, and Animal, delivered at Laval University, 1965).

Annotation: This is the classic work stating the principles and practice of what Voisin called "rational grazing," a grazing system based on plant-growth cycles and animal health and habits. Moving animals from pasture to pasture was not a new idea in the late 1950s; however, this work, focused on the science of the "meeting of cow and grass." His "laws of rational grazing" laid the foundation for "rotational grazing," "intensive grazing," "management intensive grazing," "pasture-based production" and related systems. MVG

Cited in: Coleman (1976); Merrill (1983)

1958

Krasil'nikov, Nikolai Aleksandrovich, 1896-

Soil Microorganisms and Higher Plants

Moscow: Academy of Sciences of the USSR, 1961 (English edition). 474p. Translated by Y. Halperin. Jerusalem, Israel Program for Scientific Translations, 1961. Bibliography, p. 413-474.

NAL Call no: QR111 K723 DNAr

Full-text: Soil and Health Library, Steve Solomon, http://www.soilandhealth.org/01aglibrary/01principles.html (accessed Jan. 1, 2007)

Annotation: Originally published in Russian in 1958, this text and the lengthy bibliographies provide references to little known Soviet soil scientists and their work. The author presents "basic information on the structure, development, variability and classification of bacteria, actinomycetes and fungi in the light of recent scientific achievements, as well as information on the importance of microorganisms in plant nutrition, the role of micro-activities in the complementary nutrition of plants, the effect of microbes on the vitamin content of plants, their importance in plant development and their influence on soil fertility." Table of Contents. MVG

1960

Hills, Lawrence Donegan, 1911- 1991

Down to Earth: Fruit and Vegetable Growing

London: Faber and Faber, 1960. 192p. Includes index.

NAL Call no: 93.5 H55

Other works by this author: *Russian Comfrey: A Hundred Tons an Acre of Stock Feed or Compost for Farm, Garden or Smallholding* (1953) (full-text: Soil and Health Library, Steve Solomon, http://www.soilandhealth.

org/01aglibrary/01principles.html) (accessed Apr. 23, 2007); *Composting for the Tropics; Written from the Experience of our Overseas Members for the Gardeners and Farmers of all Hot Countries* (1963); *Fertility without Fertilizers: A Basic Approach to Organic Gardening* (1977); *Fertility Gardening (1980); Fighting Like the Flowers: An Autobiography* (1989).

Annotation: Journalist and founder of the Henry Doubleday Research Association in Britain (in 1954), Hills was also a horticulturist and very knowledgeable about organic production. MVG

Cited in: Coleman (1976)

1962

Carson, Rachel, 1907-1964

Silent Spring

Boston: Houghton Mifflin, 1962. 368p. Drawings by Lois and Louis Darling. Other editions: 25th anniversary edition, 1987.

NAL Call no: 423 C23

Other works by this author: *The Sea Around Us* (1951).

Annotation: Ahead of her time in many respects, Carson's revelations about the ecological impacts of pesticide use emphasized the interconnectedness of all life. The impact her work helped launch a widely supported environmental movement in the U.S. and worldwide that continues to influence scientific research and policy. Carson's work also proved to be a turning point for interest in modern organic farming. MVG

Cited in: Conford (2001); Esbjornson (1992); Heckman (2006); Kirschenmann (2004); Kuepper, Gegner (2004); Madden (1998); Merrill (1983)

1964

DeBach, Paul, editor

Biological Control of Insect Pests and Weeds

New York: Reinhold, 1964. xxiv, 844p. Assistant editor, Evert I. Schlinger. Bibliography, p. 715-815. Other editions: 2nd edition, 1991.

NAL Call no: 423 D354

Other works by DeBach: *Biological Control by Natural Enemies* (1974).

Annotation: Classical biological control based on identifying and importing exotic natural enemies (parasites, predators, or pathogens) and utilizing them for long-term control of target exotic pests, insect or weed has been actively practiced in the U.S. since the late 1900s. DeBach's definition broadened the scope of the term to include more than just the use of natural enemies to "actions of parasites, predators and pathogens in maintaining another organism's density at a longer average than would occur in their absence." MVG

Cited in: Merrill (1983)

1964

Hunter, Beatrice Trum, 1918-

Gardening without Poisons

Boston: Houghton-Mifflin, 1964. 318p. Includes index. List of sources; List of suppliers of materials. Appendices. List of Organizations interested in gardening without poisons. Other editions: 2nd edition, 1971.

NAL Call no: SB975 H91

Other works by this author: *The Natural Foods Primer: Help for the Bewildered Beginner* (1972); *The Mirage of Safety: Food Additives and Federal Policy* (1982).

Annotation: One of the early works on biological control of insect pests, a 1971 postscript by the author cites research being done by the USDA. JPG

Cited in: Coleman (1976); Rateaver (1973)

1964

Poirot, Eugene M., 1899-1988

Our Margin of Life

New York: Vantage Press, 1964. 159p. Introduction by William A. Albrecht. Reprinted by Acres U.S.A., 1978.

NAL Call no: S624 AlP6

Annotation: A Missouri farmer/author writes seriously about agriculture, the importance of the soil and a philosophy based on his observations and beliefs. JPG

Cited in: Harwood (1983); Merrill (1983)

1968

Boucher, Jean

Precis Scientifique et Pratique de Culture Biologique: Methode Lemaire-Boucher

Angers: Agriculture et Vie, 1968. 281p. In French. Translated title: Scientific and Practical Precise of Biological Culture; Lemaire-Boucher Method. Bibliography, p. 281.

NAL Call no: S517.F8B6 1968

Other works by this author: *Une Veritable Agriculture Biologique* (1992)

Annotation: The system espoused by Dr. Raoul Lemaire (1884-1972), natural grain merchant and researcher, and Dr. Boucher, soil scientist, incorporates organic and agroecological concepts, with a focus on magnesium (per the human nutrition research of Dr. Pierre Delbet) and the use of calcified seaweed as soil amendment. MVG. See also: Delbet, 1945.

Cited in: Harwood (1983)

1968

Rusch, Hans Peter, 1906-1977

Bodenfruchtbarkeit; eine Studie Biologischen Denkens

Heidelberg: K. F. Haug, 1968. 243p. In German. Translated title: Soil Fertility. Bibliography, p. 241-243.

NAL Call no: S598.R8

Annotation: Rusch, a German soil microbiologist, teamed with Swiss organic-biological farmers Hans Müller (1891-1988) and his wife, Maria Müller (1894-1969) in the early 1950s, to lay the foundation for organic agriculture in German-speaking countries. Müller coined the German term, "organisch-biologischer landbau" (organic-biological farming) in 1949. MVG

1971

Borlaug, Norman Ernest, 1914-

Mankind and Civilization at Another Crossroad

Rome: Food and Agriculture Organization of the United Nations, 1971. 73p. "1971 McDougall Memorial Lecture."

NAL Call no: HD1417.B67 1971

Other works by this author: *Exploratory Genetic Research Undertaken by CIMMYT in Maize, Wheat, Barley and Triticale,* with E.W. Sprague (1979);

Wheat in the Third World, with Haldore Hanson and R. Glenn Anderson (1982); *Land Use, Food, Energy and Recreation* (1983); *Vetiver Grass: A Thin Green Line Against Erosion*, with Rattan Lal, David Pimentel, Hugh Popenoe, Noel Vietmeyer (1994); *The Green Revolution: An Unfinished Agenda* (2004) (full-text: of lecture online at http://www.fao.org/docrep/meeting/008/J3205e/j3205e00.htm) (accessed Apr. 23, 2007).

Annotation: The author was a central figure in the "Green Revolution" of the 1950s and 1960s and was the winner of the 1970 Nobel Prize for Peace. He helped develop high-yielding varieties of wheat, rice and corn for use in developing countries. These crops increased food production tremendously in some places, especially in Asia, while requiring extensive fertilizer and pesticide use. This lecture confronts the environmental critics of the Green Revolution. Borlaug's views are emblematic of the divide between technologically- and environmentally-driven concepts of sustainable agricultural systems. MVG

Cited in: Kirschenmann (2004)

1971

Lappé, Frances Moore

Diet for a Small Planet

New York: A Friends of the Earth/Ballantine Book, 1971. xiv, 301p. Illustrated by Kathleen Zimmerman and Ralph Iwamoto. Other editions: Revised and updated edition, 1982.

NAL Call no: TX838.L3

Other works by this author: *Food First: Beyond the Myth of Scarcity*, with Joseph Collins (1977, revised, 1979); *What Can We Do, a Food, Land, Hunger Action Guide*, with William Valentine (1980); *Hope's Edge: The Next Diet for a Small Planet*, with Anna Lappé (2002).

Annotation: Lappe's groundbreaking book alerted people to the health and environmental impacts of a meat- and processed food-oriented diet. Her dietary recommendations and their rationale, and accompanying recipes, are timeless. MVG

1971

Logsdon, Gene

Two Acre Eden

Garden City NY: Doubleday, 1971. 216p. Bibliography, p. 212-216. Other editions: Revised edition, 1980.

NAL Call no: SB455.L58

Other works by this author: *Homesteading: How to Find New Independence on the Land* (1973); *Small-scale Grain Raising* (1977); *Getting Food from Water: A Guide to Background Aquaculture* (1978); *The Contrary Farmer* (1993); *All Flesh is Grass: The Pleasures and Promises of Pasture Farming* (2004).

Annotation: Focused on rural living and growing food in sustainable ways, this prolific author has provided practical information, stories and humorous commentary to countless back-to-the-landers, gardeners and farmers over the years. MVG

1971

Rodale, Robert and Glenn F. Johns, 1930-1990

The Basic Book of Organic Gardening

New York: Ballantine, 1971. 377p. Based on material which has appeared in Organic Gardening Magazine.

NAL Call no: S605.5.R6

Other works by Robert Rodale: *The Challenge of Earthworm Research*, editor (1961); *The Organic Way to Mulching* (1972); *The Prevention System for Better Health*, with J.I. Rodale (1974); *The Cornucopia Papers* (1982); *Save Three Lives: A Plan for Famine Prevention* (1991). See also: Oral History Interview with Mr. Robert Rodale, AFSIC (1989) (NAL Call #: Videocassette no.670).

Annotation: Robert Rodale, journalist, publisher and life-long proponent of regenerative agriculture, expanded on the work his father J.I. Rodale started in the 1930s. In addition to leading the influential Rodale Publishing organization (including Organic Gardening Magazine and Prevention Magazine), he expanded The Rodale Institute's education, training and research programs. A third generation of Rodales currently carries on this work. MVG

Cited in: Kirschenmann (2004)

1972

Meadows, Donella, 1941-2001

Limits to Growth: A Report for the Club of Rome's Project on the Predicament of Mankind

New York: Universe Books, 1972. 205p. Bibliography, p. 198-205.

NAL Call no: HB871.L5

Other works by this author: *The Electric Oracle: Computer Models and Social Decisions*, with J.M. Robinson (1985); *Global Citizen* (1991); *Beyond the Limits: Confronting Global Collapse, Envisioning a Sustainable Future*, with Dennis Meadows and Jorgen Randers (1993); *Limits to Growth—The 30 Year Update*, with Jorgen Randers and Dennis Meadows (2004).

Annotation: This study, utilizing computer modeling, examined global economic, population and environmental trends. It was an influential work in the then new debate on earth's "carrying capacity." Meadows paid special attention to sustainable food production systems, developing an ecovillage and an organic farm of her own. She also created innovative information sharing initiatives and a widely-read weekly news column, Global Citizen. MVG

1973

Hightower, Jim and Agribusiness Accountability Project Task Force on the Land Grant College Complex

Hard Tomatoes, Hard Times: A Report of the Agribusiness Accountability Project on the Failure of America's Land Grant College Complex

Cambridge MA: Schenkman, 1973. 268p. Includes bibliographical references. Reissued with "selected additional views of the problems and prospects of American agriculture in the late seventies," 1978.

NAL Call no: S533.H523 1973

Other works by this author: *Eat your Heart Out: Food Profiteering in America* (1975).

Annotation: This is a biting criticism of Land Grant institution research policy and extension activities. Hightower cites their lack of attention to the interests of small farms and farmers, rural communities and farm workers in favor of agribusiness, large-scale farms and mechanized, production-oriented technology. MVG

1973

Schumacher, Ernst Friedrich, 1911-1977

Small is Beautiful; Economics as if People Mattered

New York: Harper and Row, 1973. 290p. Includes bibliographical references.

NAL Call no: HB171.S384

Other works by this author: *Think about Land* (1974).

Annotation: E.F. Schumacher was very concerned with "The Proper Use of Land" (Part 2 of this book) and the scale of agricultural production. He became involved in organic farming, serving as president of the UK's Soil Association. He espoused conservation in all things including food production and energy use, and advocated the use of "appropriate technology" whenever possible. MVG

Cited in: Scofield (1986)

1974

Jeavons, John

How to Grow More Vegetables Than You Ever Thought Possible on Less Land Than You Can Imagine

Palo Alto CA: Ten Speed Press, 1979. 116p. Bibliography. Other editions: First edition of this book was published in 1974; most recent edition: 7th edition, Ten Speed Press, 2006.

NAL Call no: SB320.6.J43

Other works by this author: *Lazy-bed Gardening: The Quick and Dirty Guide*, with Carol Cox (1993); *The Sustainable Vegetable Garden: A Backyard Guide to Healthy Soil and Higher Yields*, with Carol Cox (1999).

Annotation: Jeavons, a former systems analyst, writes a "primer on the lifegiving biodynamic/French intensive method of organic agriculture." He is involved with "Ecology Action" in Palo Alto, California. JPG. [Ecology Action is now based in Willets CA. MVG]

Cited in: Harwood (1983); Harwood (1990); Kuepper, Gegner (2004)

1975

Lockeretz, William, Robert Klepper, Barry Commoner, Michael Gertler, Sarah Fast, Daniel O'Leary and Roger Blobaum

A Comparison of the Production, Economic Returns, and Energy-intensiveness of Corn Belt Farms that Do and Do Not Use Inorganic Fertilizers and Pesticides

St. Louis MO: Washington University, Center for the Biology of Natural Systems, 1975. 62p. Includes bibliographical references. (CBNS Report AE 4)

NAL Call no: S605.5.C59 1975

Other works by Lockeretz: *Agricultural Resources Consumed in Beef Production* (1975); *Linkages Between Farming and Other Economic Activities in Agriculturally-dependent Areas* (1988); *Agricultural Research Alternatives,* with Molly D. Anderson. (1993); *Visions of American Agriculture,* editor (2000); *Ecolabels and the Greening of the Food Market,* editor (2003). See also: Oral History Interview with William Lockeretz, AFSIC (1991) (NAL Call #: Videocassette no.1217).

Annotation: This report was one of the first U.S. studies analyzing the economics of organic farming systems. Lockeretz and other agricultural economists of the 1970s led the way to new research on sustainable systems' economic performance, crop yield, energy inputs, cost analysis, conventional vs. organic comparisons and more. MVG

Cited in: Kuepper, Gegner (2004); Harwood (1983)

1976

Boeringa, Rob, editor

Alternative Methods of Agriculture

Amsterdam: Elsevier, 1980. 199p. Translation of parts of the original Dutch report, Alternatieve Landbouwmethoden, prepared in 1976 by the Committee for Research into Biological Methods of Agriculture. References.

NAL Call no: S60l.D4 Vol.10

Annotation: A translation of selections of the famous Dutch report of 1976, this work includes descriptions of each school of ecological agriculture and a literature review on techniques, yields, food quality, impact on the environment and research recommendations. JPG

Cited in: Harwood (1983); Merrill (1983)

1976

Koepf, Herbert H., Bo D. Pettersson and Wolfgang Schaumann

Biodynamic Agriculture: An Introduction/ Biologische Landwirtschaft

Spring Valley NY: Anthroposophic Press, 1976. x, 429p. Translated from the German, Biologische Landwirtschaft (1974). Index. Bibliography, p. 404-416.

NAL Call no: S605.5.K59

Full-text: Soil and Health Library, Steve Solomon, http://www.soilandhealth.org/01aglibrary/01principles.html (accessed Jan. 1, 2007)

Other works by Koepf: *What is Bio-dynamic Agriculture?* (translated by W. Brinton and M. Spock, 1979); *Bio-dynamic Sprays* (1988); *The Biodynamic Farm: Agriculture in the Service of the Earth and Humanity,* with R. Shouldice and W. Goldstein (1989); *Soil Fertility in Sustainable Low Input Farming,* edited by J. L. Ruhnau (1992).

Annotation: This work was originally published in German so the bibliography is of unusual interest. JPG

Cited in: Blum (1993); Harwood (1983); Harwood (1990); Merrill (1983)

1976

Merrill, Richard, editor, 1941-

Radical Agriculture

New York: Harper and Row, 1976. xix, 459p. Includes bibliographical references and index.

NAL Call no: S441.R26

Annotation: One of the earliest collections of essays focusing on alternative and sustainable agriculture, this compilation examines a variety of farm topics including rural and labor issues, aquaculture, renewable energy, cities and farms, land reform and more. Essays come from highly regarded agricultural thinkers including Michael Perelman, Wendell Berry, Jim Hightower, Jerome Goldstein and Helga and William Olkowski. MVG

Cited in: Harwood (1983)

1976

United States Congress

The Farmer-to-Consumer Direct Marketing Act of 1976

Washington DC: United States Congress, 1976.

NAL Call no: aHD9003.N37

Full-text: United States Code, Office of the Law Revision Counsel, http://uscode.house.gov/download/pls/07C63.txt (accessed Jan. 1, 2007)

See also: National Farmers' Market Directory, 2006 (full-text: USDA, Agricultural Marketing Service, http://www.ams.usda.gov/farmersmarkets/map.htm) (accessed May 7, 2007).

Annotation: Public Law 94–463, October 8, 1976, "An act to encourage the direct marketing of agricultural commodities from farmers to consumers," heralded a rapid increase in the number of U.S. farmers' markets. Provisions of this act, along with strong consumer demand, continue to support growth of local and regional markets and other direct marketing activities. The U.S. Department of Agriculture's National Farmers' Market Directory, first compiled in 1994, listed 4,385 U.S. farmers markets in 2006 (up from 1,755 markets in 1994). USDA Releases New Farmers Market Statistics, AMS Program Announcement, Dec. 5, 2006 (http://www.ams.usda.gov/tmd/MSB/PRFarmersMarketStatistics.pdf) (accessed May 7, 2007). MVG

1977

Berry, Wendell, 1934-

The Unsettling of America: Culture and Agriculture

San Francisco: Sierra Club Books, 1977. 228p. Chapter notes. Other editions: 2nd edition, Sierra Club Books, 1986. 3rd edition, Sierra Club Books, 1996.

NAL Call no: HD1761.B47

Other works by this author: *The Gift of Good Land: Further Essays, Cultural and Agricultural* (1981); *What are People For? Essays* (1990); *Citizenship Papers* (2003).

Annotation: A review/criticism on agricultural policy today, this work presents the author's position: "the care of the earth is our most ancient and most worthy and, after all, our most pleasing responsibility." JPG

Cited in: Esbjornson (1992); Harwood (1983); Harwood (1990); Kirschenmann (2004)

1977

Dideriksen, Raymond Ivan, Allen Ray Hidlebaugh and Keith O. Schmude

Potential Cropland Study

Washington DC: Soil Conservation Service, U.S. Dept. of Agriculture, 1977. 104p. Chiefly tables. (Statistical Bulletin, 578)

NAL Call no: 1 Ag84St no.578

Annotation: The 1970s saw a growing concern with land-use issues and with loss of farmland in particular. This study, done at the request of Earl Butz, U.S. Secretary of Agriculture, was one of the first to document losses of farmland to urban development. During 1967-1975, "nearly 2.1 million acres each year were converted to urban and built-up areas. About 30 percent of the land converted to urban and built-up areas each year comes from cropland." This publication helped solidify government and conservation group support to preserve agricultural lands. MVG

Cited in: Lehman (1993)

1977

Lovins, Amory B. and Friends of the Earth International

Soft Energy Paths: Toward a Durable Peace

San Francisco: Friends of the Earth International, 1977. xx, 231p. Includes bibliographical references and index.

NAL Call no: TJ163.2.L678

Other works by this author: *Energy Unbound: A Fable for America's Future,* with L Hunter Lovins and Seth Zuckerman (1986); *Natural Capitalism: Creating the Next Industrial Revolution,* with Paul Hawken and L. Hunter Lovins (1999); *Small Is Profitable: The Hidden Economic Benefits of Making Electrical Resources the Right Size* (2003); *Winning the Oil Endgame: Innovation for Profit, Jobs and Security,* with E. Kyle Datta, Odd-Even Bustnes, Jonathan G. Koomey and Nathan J. Glasgow (2005) (see Online books by Lovins at http://onlinebooks.library.upenn.edu/webbin/book/lookupname?key=Lovins%2C%20Amory%20B.%2C%201947-) (accessed Apr. 23, 2007).

Annotation: Amory Lovins and the work of the Rocky Mountain Institute where he is CEO, have influenced ideas about energy production and use in agriculture. His "soft energy path" advocates an approach that combines energy conservation and efficiency; use of renewable sources of energy; and local, small-scale generation and use of energy. MVG

1978

Besson, J. M. and H. Vogtmann, editors

Towards a Sustainable Agriculture

Aarau, Switzerland: Wirz; International Federation of Organic Agriculture Movements, 1978. 243p. Papers given at an international conference held in Sissach, Switzerland, 1977. References at ends of articles.

NAL Call no: S605.5.T68

Annotation: These are proceedings of the lst conference of the IFOAM, in French, German and English. All IFOAM proceedings are recommended. JPG. [A catalog of available IFOAM published proceedings may be found at http://shop.ifoam.org/bookstore/ (accessed May 21, 2007)]

Cited in: Harwood (1983); Merrill (1983)

1978

Brown, Lester Russell, 1934-

The Twenty-ninth Day: Accommodating Human Needs and Numbers to the Earth's Resources

New York: Norton, 1978. xiii, 363p.

NAL Call no: GF41.B76

Other works by this author: *American Agriculture in a Hungry World* (1967); *The Social Impact of the Green Revolution* (1971); *Who will Feed China? Wake-up Call for a Small Planet* (1995); *The Agricultural Link: How Environmental Deterioration Could Disrupt Economic Progress* (1997); *Outgrowing the Earth: The Food Security Challenge in the Age of Falling Water Tables and Rising Temperatures* (2004).

Annotation: Mr. Brown, an agricultural economist and a pioneer in the environmental movement, founded the Worldwatch Institute in 1974. In this work, unsustainable uses of water, soil and energy resources are discussed in the context of projected population growth, inappropriate political decision-making and inadequacy of technical remedies. MVG

1978

Dahlberg, Kenneth A.

Beyond the Green Revolution: The Ecology and Politics of Global Agricultural Development

New York: Plenum Press, 1978. xiii, 256p. Includes bibliographical references and index.

NAL Call no: HD1415.D273

Other works by this author: *New Directions for Agriculture and Agricultural Research: Neglected Dimensions and Emerging Alternatives,* editor (1986).

Annotation: The "Green Revolution" of the 1960s and 1960s was characterized by innovative, high-yielding varieties of cereal crops developed for use in developing countries. These crops boosted food security while requiring extensive technology for planting, irrigation, fertilizing, pest control and harvesting. Dahlberg outlines the social and ecological costs of this approach and makes the case for more sustainable agricultural alternatives in both industrialized and developing countries. MVG. See also: Borlaug, 1971.

Cited in: Harwood (1990)

1978

Gussow, Joan Dye

The Feeding Web: Issues in Nutritional Ecology

Palo Alto CA; New York: Bull Publishing; trade distribution in U.S. by Hawthorn Books, 1978. xvi, 457p.

NAL Call no: HD9000.5 .F36

Other works by this author: *The Nutrition Debate: Sorting Out Some Answers,* with Paul Thomas (1986); *Chicken Little, Tomato Sauce and Agriculture: Who Will Produce Tomorrow's Food?* (1991); *This Organic Life: Confessions of a Suburban Homesteader* (2001).

Annotation: Dr. Gussow, a nutrition professor at Teachers College, Columbia University, was the first to use the term "nutritional ecology," and to focus on the entire food chain in terms of environmental, nutritional and economic sustainability. Today's food system includes not only production and harvesting, but storing; transporting; processing and packaging; distribution and trade; food composition; food preparation and consumption; and waste disposal. She argues that consumers need to make food choices in terms of not only health issues, but in terms of environmental and social values as well. MVG

1978

Mollison, Bill C. and David Holmgren

Permaculture 1: A Perennial Agricultural System for Human Settlements

Hobart, Australia: Environmental Psychology, University of Tasmania, 1978. vii, 128p. Bibliography, p. 126-127.

NAL Call no: S589.7.M6

Other works by Mollison: *Permaculture Two: Practical Design for Town and Country in Permanent Agriculture* (1979; updated 1999); *Permaculture: A Designers' Manual* (1988); *Introduction to Permaculture,* with Reny Mia Slay (1991).

Annotation: Australians Mollison and Holmgren co-developed the permaculture system of landscape design, a holistic system now taught and implemented throughout the world. MVG

1978

Oelhaf, Robert C., 1938-

Organic Agriculture: Economic and Ecological Comparisons with Conventional Methods

Montclair NJ: Allanheld, Osmun, and Company, 1978. 271p. Includes index. References.

NAL Call no: S605.5.066

Other works by this author: *The Economics of Organic Farming* (Ph.D. dissertation, University of MD, 1976).

Annotation: The author, trained in theology, science and engineering, looks at more than just comparative economic issues and presents a detailed study of internal and external costs and benefits of ecological and conventional agriculture. JPG

Cited in: Harwood (1983); Heckman (2006); Merrill (1983)

1978

Schwartz, James W.

A Bibliography for Small and Organic Farmers: 1920-1978

Beltsville MD: Agricultural Research, United States Department of Agriculture, 1978. Same title published in 1981 as number 11 in the series, Bibliographies and Literature of Agriculture.

NAL Call no: aZ5074 O7S3

Annotation: This bibliography cites documents that relate to organic and small-scale farming published from 1920 through 1978. USDA National Agricultural Library staff continues to compile bibliographies on topics related to sustainable, organic and alternative farming. See the Alternative Farming Systems Information Center (AFSIC) Web site for updated bibliographies on sustainable and organic topics (http://afsic.nal.usda.gov) (accessed Feb. 1, 2007). MVG

1978

United States Congress House Committee on Agriculture

Agricultural Land Retention Act: Report Together with Dissenting Views to Accompany H.R. 11122

Washington DC: Government Printing Office, 1978. (House Report - 95th Congress, 2d Session, 95-1400)

NAL Call no: KF32.A3 1978b

Annotation: The Agricultural Land Retention Act, proposed in 1977, was one of the first attempts to address the loss of farmland to urban development through legislation. It ultimately failed to pass the Congress (in 1980). Many of its provisions have since been instituted in the Farmland Protection Policy Act, the 1981 Agriculture and Food Act and in the form of new agencies and state and local legislation. Non-governmental programs and advocacy continue to keep farmland protection near the top of the sustainable agriculture agenda. MVG

Cited in: Lehman (1993)

1978

Whealy, Kent, editor

The Third Annual True Seed Exchange

Princeton MO: Kent Whealy, 1978.

NAL Call no: SB115.T5

Annotation: Whealy and his wife, Diane Ott Whealy, founded the Seed Savers Exchange in 1975. It was one of the first organizations to alert gardeners and farmers to the importance of saving "heirloom" seeds. "The genetic diversity of the world's food crops is eroding at an unprecedented and accelerating rate. The vegetables and fruits currently being lost are the result of thousands of years of adaptation and selection in diverse ecological niches around the world..." Seed Savers Web site (http://www.seedsavers.org/) (accessed Feb. 1, 2007). MVG

1979

Cox, George W. and Michael D. Atkins

Agricultural Ecology: An Analysis of World Food Production Systems

San Francisco: W. H. Freeman, 1979. 721p. Includes bibliographies and index.

NAL Call no: S589.7.C69

Annotation: Agroecology, a multi-disciplinary field established in the 1980s, owes much to this pioneering work. Studies and programs on the topic continue to point the way to sustainable agricultural systems and practices. MVG. See also: Altieri, 1983 and Gliessman, 1990.

1979

Minnich, Jerry and Marjorie Hunt, editors

Rodale Guide to Composting

Emmaus PA: Rodale Press, 1979. 405p. Includes index. Bibliography, p. 385-389.

NAL Call no: S661.M56

Annotation: The most complete book on composting at the time, this volume offered credible science and practical methodology for farmers and gardeners. MVG

Cited in: Harwood (1983); Harwood (1990)

1979

Pimentel, David and Marcia Pimentel

Food, Energy, and Society

New York: Wiley, 1979. viii, 165p. Includes index. Bibliography, p. 151-162. Most recent edition: University Press of Colorado, 1996. (Resource and Environmental Sciences Series)

NAL Call no: HD9000.6.P55 1979a

Other works by David Pimentel: *Ecological Effects of Pesticides on Nontarget Species* (1971); *World Food, Pest Losses, and the Environment,* editor (1978); *Handbook of Energy Utilization in Agriculture,* editor (1980); *CRC Handbook of Pest Management in Agriculture,* editor (1981); *Water Resources, Agriculture and the Environment* (2004) (full-text: Cornell University, http://dspace.library.cornell.edu/bitstream/1813/352/1/pimentel_report_04-1.pdf) (accessed Apr. 23, 2007).

Annotation: The Pimentels' groundbreaking work details energy use in the production of livestock, grains, fruits, vegetables, forage crops and fish; food processing and transport; and the impact of energy use on the environment. Dozens of charts and tables, as well as a lengthy reference list, are included. "Thus, the aim of this book is to explore the interdependencies of food, energy and their impacts on society. These analyses we hope will be a basis for planning and implementing policies of individuals and nations as they face the inevitable dilemma—how can everyone be fed, given the limited resources of the earth." Preface. MVG

1979

Walters, Charles Jr., 1926-; and C. J. Fenzau

An ACRES U.S.A. Primer

Raytown MO: Acres U.S.A., 1979. 465p. Glossary. Index. Field notes. Lists of Eco-suppliers. Most recent edition: Ecofarm: An Acres U.S.A. Primer, 3rd edition, 2003.

NAL Call no: S605.5.W34

Other works by this author: *The Carbon Connection, with Leonard Ridzon* (1990); *Weeds: Control without Poisons* (1991, revised 1999); *A Farmer's Guide to the Bottom Line* (2002); *Reproduction and Animal Health,* with Gearld Fry (2003).

Annotation: A "first reader" in eco-agriculture, this book is written from an organic perspective on the basis of plant and soil science, agronomy and pest control. JPG

Cited in: Harwood (1983); Kuepper, Gegner (2004); Merrill (1983)

1980–2007

THE SUSTAINABLE AGRICULTURE AND MODERN ORGANIC FARMING MOVEMENTS TAKE ROOT

The late 20th century saw great progress in sustainable agriculture practice, policy and research activity. The term "sustainable agriculture" was adopted during this period. The works below speak to the great variety of farming topics that have come to fall under the sustainability umbrella.

ESSAY COLLECTIONS AND CONFERENCE PROCEEDINGS, 1980S

A wide range of ideas and knowledge was presented in collections of research reports, conference papers and philosophical essays written by farmers, researchers and policy makers. A researched selection of compilations is listed in this section.

1980

Jackson, Wes, 1936-

New Roots for Agriculture

Lincoln NE: University of Nebraska Press, 1980. 151p. References and notes at ends of chapters. Other editions: Revised edition 1985.

NAL Call no: S441.J25

Other works by this author: *Altars of Unhewn Stone: Science and the Earth* (1987); *Becoming Native to This Place* (1994). See also: Oral History Interview with Wes Jackson, AFSIC (1990) (NAL Call #: Videocassette no.731).

Annotation: "In popular literature, sustainable agriculture generally is presented as a new phenomenon. Wes Jackson is credited with the first publication of the expression in his New Roots for Agriculture (1980) and the term didn't emerge in popular usage until the late 1980s." "A Brief History of Sustainable Agriculture," by Fred Kirschenmann, in The Networker, vol. 9, no. 2 (Science and Environmental Health Network), March 2004. The new edition (University of Nebraska Press, 1985) has a foreword by Wendell Berry and an "afterward" to its conclusion, p. 133-148. JPG

Cited in: Beeman (1993); Esbjornson (1992); Harwood (1983); Harwood (1990); Kirschenmann (2004)

1980

United States Department of Agriculture Study Team on Organic Farming

Report and Recommendations on Organic Farming

Washington DC: United States Department of Agriculture, 1980. 94p. References at ends of chapters. Study team members: Dr. Robert I. Papendick (Coordinator and Chairman); Dr. Larry L. Boersma; Daniel Calacicco; Joanne M. Kla; Dr. Charles A. Kraenzle; Dr. Paul B. Marsh; Dr. Arthur S. Newman; Dr. James F. Parr; Dr. James B. Swan; Dr. I Garth Youngberg.

NAL Call no: aS605.5 U52

Full-text: Alternative Farming Systems Information Center, http://afsic.nal.usda.gov/afsic/pubs/USDAOrgFarmRpt.pdf (accessed Jan. 1, 2007)

See also: Oral History Interview with I. Garth Youngberg, AFSIC (1991) (NAL Call #: Videocassette no.1128).

Annotation: Compiled with the assistance of the Rodale Press survey of The New Farm readers, this study was "conducted to learn more about the potential contributions of organic farming as a system for the production of food and fiber." A first for the U.S., it is a general overview of the status of organic agriculture in the United States: methods, implications for environmental and food quality, economic assessment and research recommendations. JPG

Cited in: Harwood (1983); Heckman (2006); Kuepper, Gegner (2004); Madden (1998); Merrill (1983)

1981

Cornucopia Project

Empty Breadbasket? The Coming Challenge to America's Food Supply and What We Can Do About It: A Study of the U.S. Food System

Emmaus PA: Rodale Press, 1981. 189p. Bibliography, p. 148-175.

NAL Call no: HD9005.E56

Works done in collaboration with The Cornucopia Project: *The New Jersey Food System: A Harvest of Doubt for the Garden State: Working Draft, a Study of the Food System of New Jersey*, by Phyllis Swackhamer (1980); *The State of Your Food: A Manual for State Food Systems Analysis* (1981); *The Pennsylvania Food System: Crash or Self-reliance? A Study of the Food System of Pennsylvania* (1981); *The New York State Food System: Growing Closer to Home, a Study of the Food System of New York State*, by Patricia Messing (1981); *The Maine Food System: A Time for Change: A Study of the Food System of Maine*, by Raymond Wirth (1981); *The Indiana Food System: Sustainable or in Jeopardy? A Study of the Food System in Indiana*, by Roberta Wysong (1982); *The Cornucopia Papers*, by Robert Rodale (1982); *Regenerating the Food System: A Food and Agricultural Policy for the United States; A Framework for Discussion* (1984); *The South Carolina Food System: Does It Have a Future? A Study of the South Carolina Food System*, by John Madera (1985).

Annotation: This report documents research and recommendation from the Rodale Institute's Cornucopia Project. It outlines methods to assess food self-sufficiency and vulnerabilities to the food system, especially at the state level. It suggests how consumers, farmers and the food industry could cooperate in developing a secure, affordable and ecologically sustainable food supply. MVG

1981

Jeske, Walter E., editor
Economics, Ethics, Ecology: Roots of Productive Conservation: Based on Material Presented at the 35th Annual Meeting of the Soil Conservation Society of America, August 4-August 6, 1980, Dearborn, Michigan
Ankeny IA: Soil Conservation Society of America, 1981. xiv, 454p. Includes bibliographies.
NAL Call no: S912.E28
Annotation: Papers from this early ecologically-oriented conference address land and water management, energy issues, and education and values in natural resource conservation efforts. MVG
Cited in: Harwood (1990)

1981

Stonehouse, Bernard, editor
Biological Husbandry: A Scientific Approach to Organic Farming

London; Boston: Butterworths, 1981. xiii, 352p. Includes papers presented in the Proceedings of the 1st International Institute of Biological Husbandry Symposium held at Wye College, Ashford, Kent, on August 26-30, 1980. Includes bibliographical references and index.
NAL Call no: S605.5.S7
Annotation: This includes "most of the papers" presented at the lst international symposium of the International Institute of Biological Husbandry, 1980. Sections include: soil structure, flora and fauna, agricultural methods, biological husbandry in the tropics, systems of agriculture and comparative studies. JPG
Cited in: Harwood (1983); Merrill (1983)

1982

Hill, Stuart B. and Pierre Ott, editors

Basic Techniques in Ecological Farming: Papers Presented at the 2nd International Conference held by the IFOAM, Montreal, October 1-5, 1978

Basel: Berkhauser Verlag, 1982. 366p.

NAL Call no: S605.5.B39

Annotation: This volume and other available proceedings of IFOAM (International Federation of Organic Agriculture Movements) conferences are highly recommended. JPG. See also: Beeson, 1978.

Cited in: Harwood (1983); Merrill (1983)

1983

Altieri, Miguel

Agroecology; the Scientific Basis of Alternative Agriculture

Berkeley: University of California, 1983. 173p.

NAL Call no: S589.7 A4

Other works by this author: *Experiences in Success: Case Studies in Growing Enough Food through Regenerative Agriculture,* with Kenneth Tull and Michael Sands (1987); *Biodiversity and Pest Management in Agroecosystems,* with Clara I. Nicholls (1994; revised edition, 2003); *The Potential of Agroecology to Combat Hunger in the Developing World,* with Peter Rosset, and Lori Ann Thrupp (1998); *Genetic Engineering in Agriculture: The Myths, Environmental Risks, and Alternatives* (2001, revised edition, 2004).

Annotation: A review of temperate and tropical agroecology, theory and practice, this work includes chapters on the design of sustainable systems, traditional peasant agriculture, polyculture, tree cropping, live mulches, minimum tillage and pest control. JPG

Cited in: Harwood (1990); Kirschenmann (2004)

1983

Batie, Sandra S.

Soil Erosion: Crisis in America's Croplands?

Washington DC: Conservation Foundation, 1983. xv, 136p. Includes bibliographies and index.

NAL Call no: S624.A1B33

Other works by this author: *Emerging Issues in Water Management and Policy*, editor, with J. Paxton Marshall (1983); *Improving Land Use Policy Analysis in the Southeast: Lessons from Virginia's Agricultural and Forestal District Act*, with E. Jane Luzar (1986); *Economic and Legal Analysis of Strategies for Managing Agricultural Pollution of Ground Water*, with Randall A. Kramer, William E. Cox (1989); *Green Payments as Foreshadowed by EQIP* (1998); *The Economics of Agri-environmental Policy*, editor, with Richard D. Horan. (2004).

Annotation: This effective presentation of statistics regarding soil loss reminded the agricultural community that soil loss continued to be a major problem in U.S. decades after the Dust Bowl. MVG

Cited in: Esbjornson (1992)

1983

Duke, James A., 1929-

Handbook of Energy Crops

Lafayette IN: Purdue University, Center for New Crops and Plants Products, 1983. Includes bibliographical references and index.

NAL Call no: SB288 .D85 1983

Full-text: Center for New Crops and Plants Products, http://newcrop.hort.purdue.edu/newcrop/duke%5Fenergy/dukeindex.html (accessed Jan. 1, 2007)

Other works by this author: *Handbook of Medicinal Herbs* (1985, 2nd edition 2002); *CRC Handbook of Agricultural Energy Potential of Developing Countries*

(1987); *A Field Guide to Medicinal Plants: Eastern and Central North America*, with Steven Foster (1990); *Handbook of Phytochemical Constituents of GRAS Herbs and Other Economic Plants* (1992); *CRC Handbook of Alternative Cash Crops*, with Judith L. du Cellier (1993); *Handbook of Legumes of World Economic Importance*, editor (2002). See also: Oral History Interview with Dr. James A. Duke, AFSIC (1988) (NAL Call #: Videocassette no.629).

Annotation: Dr. Duke, a USDA Agricultural Research Service (ARS) botanist for many years, researched and compiled dozens of texts about unusual plants, energy crops, medicinal and culinary herbs and ethnobotany. His work has pointed the way to many viable alternative crops and enterprises for U.S. farmers, researchers and businesses. MVG

1983

Knorr, Dietrich, editor

Sustainable Food Systems

Westport CT: AVI Publishing, 1983. xiv, 416p. Includes bibliographies and index.

NAL Call no: TP370.5.S94

Annotation: This compilation is one of the first that combined a range of disciplines related to sustainable agriculture - agricultural and food production practices, rural and urban food production issues, food processing and distribution topics and nutritional and food quality factors related to ecologically grown foods. MVG

1984

Institute for Alternative Agriculture

Alternative Agriculture: An Introduction and Overview: Symposium Proceedings, March 1984, Washington, D.C.

Greenbelt MD: Institute for Alternative Agriculture, 1984. ii, 49p. Includes bibliographies. An Institute publications list, including links to selected full-text reports, is maintained at Winrock International (http://www.winrock.org/wallace/publications.asp?BU=9064) (accessed Apr. 23, 20007).

NAL Call no: S605.5.A47

Annotation: Founded in 1983, the Institute, later named the Henry A. Wallace Institute for Alternative Agriculture and the Henry A. Wallace Center for Agricultural and Environmental Policy at Winrock International, provided a forum, research program and advocacy platform for policy issues pertinent to sustainable agricultural production. These are Proceedings from the Institute's 1st annual scientific symposium held in March 1984. Researchers, including Institute founder, Garth Youngberg, presented papers on production strategies, environmental benefits, research needs and public policy. MVG

1984

Jackson, Wes, Wendell Berry and Bruce Colman, editors

Meeting the Expectations of the Land: Essays in Sustainable Agriculture and Stewardship

San Francisco: North Point Press, 1984. xvi, 250p. Bibliography, p. 231-247.

NAL Call no: S441.M4

Annotation: This outstanding collection includes, in addition to works by each editor, essays by Marty Bender, Dana Jackson, Gene Logsdon, Donald Worster, Marty Strange, John Todd, Gary Paul Nabhan, Stephen Gliessman, Angus Wright, Jennie Gerard and Sharon Johnson and Gary Snyder. MVG

1984

Kral, David M., editor

Organic Farming: Current Technology and its Role in a Sustainable Agriculture

Madison WI: American Society of Agronomy (ASA), Crop Science Society of America (CSSA) and Soil Science Society of America (SSSA), 1984. vii, 192p. Proceedings of a Symposium, Sponsored by Division S3, S4, S6, S8 and A5 of the American Society of Agronomy, Crop Science Society of America and the Soil Science Society of America in Atlanta GA, Nov. 9 -Dec. 3, 1981. D.F. Bezdicek, chairman. (ASA Special Publication, 46)

NAL Call no: 64.9 Am3 no.46

Annotation: This presentation made at the 1981 Symposium represents the first documented dialog about organic farming at a national "Tri-Society" meeting. MVG

Cited in: Heckman (2006); Kuepper, Gegner (2004)

1984

Lowrance, Richard, Benjamin R. Stinner and Garfield J. House, editors

Agricultural Ecosystems; Unifying Concepts

New York: John Wiley, 1984. 233p. Includes index. References at ends of chapters.

NAL Call no: S589.7 A36

Annotation: Written by 21 of the foremost thinkers on the subject of agroecosystems, the majority of the chapters were presented initially at a 1982 symposium held during a meeting of the Ecological Society of America. JPG

1984

Postel, Sandra

Water: Rethinking Management in an Age of Scarcity

Washington DC: Worldwatch Institute, 1984. 65p. "This paper will appear as the chapter 'Managing Freshwater Supplies' in State of the World 1985," p. 5. Includes bibliographical references, p. 55-65. (Worldwatch Paper, 62)

NAL Call no: HD1691.P67

Other works by this author: *Water for Agriculture: Facing the Limits* (1989); *Last Oasis: Facing Water Scarcity* (1992, revised in 1997); *Rivers for Life: Managing Water for People and Nature,* with Brian Richter (2003).

Annotation: Water availability and consumption in relation to agricultural production has often been overlooked. This data-packed volume, with extensive references, was one of the first to define the many issues related to sustainable water management. MVG

1984

Todd, John and Nancy Jack Todd

Bioshelters, Ocean Arks, City Farming: Ecology as the Basis of Design

San Francisco: Sierra Club Books, 1984. 210p. Includes index. Bibliography, p. 169-176.

NAL Call no: GF50.T6

Other works by these authors: *The Village as Solar Ecology: Proceedings of the New Alchemy/Threshold Generic Design Conference*, editors (1980); *From Eco-cities to*

Living Machines: Principles of Ecological Design (1994); *A Safe and Sustainable World: The Promise of Ecological Design*, by Nancy Jack Todd (2005).

Annotation: The Todds' innovative work with the New Alchemy Institute in Massachusetts suggested practical ways for looking at food production and waste recycling within a totally integrated, self-contained living system. MVG

1984

Worldwatch Institute

State of the World: A Worldwatch Institute Report on Progress Toward a Sustainable Society

New York: Norton, 1984. Published annually beginning in 1984.

NAL Call no: HC59.S73

Full-text: Worldwatch Institute Information page, http://www.worldwatch.org/taxonomy/term/38 (accessed Jan. 1, 2007)

Annotation: This volume is the first in a series of annual presentations of data, analysis and case studies related to environmental and social sustainability - population growth, energy use, soil and water conservation, forest resources, recycling materials, food security and more. The 2007 edition is entitled, Our Urban Future and includes a chapter on "Farming in Cities." MVG

Cited in: Harwood (1990)

1985

Edens, Thomas, Cynthia Fridgen and Susan L. Battenfield, editors

Sustainable Agriculture and Integrated Farming Systems: 1984 Conference Proceedings

East Lansing MI: Michigan State University Press, 1985. iv, 344p. 29+ papers. Includes bibliographies.

NAL Call no: S441.S8

Annotation: Many of sustainable agriculture's U.S. and European pioneers made presentations at this ground-breaking conference: Gordon Douglass, Engelhard Boehncke, Herbert Koepf, Herman Koenig, Eliot Coleman, Stephen Gliessman, Richard Harwood, Helene Hollander, Nicolas Lampkin, William Lockeretz, Robert H. Miller, George Bird, Pieter Vereijken, Hartmut Vogtmann, Stephen Kaffka, Miguel Altieri, Terry Cacek, Gunter Kahnt, Dean Haynes,

Kenneth Dahlberg, Maynard Kaufman, Robert Bealer, Michael Dalecki, Philip Shepard, Edwin French, Dwight Schmidt, J.Patrick Madden, Gerhard Plakholm, Lawrence Woodward, Frederick Buttel, Garth Youngberg, Harold Breimyer, Robert Rodale. Rodale states in his keynote address, "By marching forward under the banner of sustainability we are, in effect, continuing to hamper ourselves by not accepting a challenging enough goal. I am not against the word sustainable, rather I favor regenerative agriculture." MVG

1985

Francis, Charles A. and Richard R. Harwood

Enough Food: Achieving Food Security through Regenerative Agriculture

Emmaus PA: Rodale Institute, 1985. 20p.

NAL Call no: S605.5.F73

Other works by Francis: *Multiple Cropping Systems* (1986); *Sustainable Agriculture in the Midwest: North Central Regional Conference*, editor (1988); *Sustainable Agriculture in Temperate Zones*, editor (1990) See also: Oral History Interview with Charles A. Francis, AFSIC (1990) (NAL Call #: Videocassette no.876). Other works by Harwood: Small Farm Development: Understanding and Improving Farming Systems in the Humid Tropics (1979); Research Agenda for the Transition to a Regenerative Food System, with J. Patrick Madden (1982); Research Towards Integrated Natural Resources Management: Examples of Research Problems, Approaches and Partnerships in Action in the CGIAR, editor (2003).

Annotation: Authored by two influential sustainable agriculture researchers and educators, this document is based on work done at the Rodale Institute. MVG

1985

Fukuoka, Masanobu, 1913-

The Natural Way of Farming: The Theory and Practice of Green Philosophy

Tokyo: Japan Publications, 1985. 280p. Includes index. Translated from the Japanese, Shizen Noho, by Frederic P. Metreaud. Other editions: Revised edition, Japan Publications, 1987.

NAL Call no: S606.6 F72

Full-text: Soil and Health Library, Steve Solomon, http://www.soilandhealth.org/01aglibrary/01principles.html (accessed Jan. 1, 2007)

Other works by this author: *The One Straw Revolution* (1978); *The Road Back to Nature: Regaining the Paradise Lost* (1988).

Annotation: Fukuoka, author of The One Straw Revolution, reiterates his five major principles: no tillage, no fertilizer, no pesticides, no weeding and no pruning. His book deals almost exclusively with farming in Japan, but his message can be viewed from a universal perspective. JPG

Cited in: Harwood (1983)

1985

Nabhan, Gary Paul

Gathering the Desert

Tucson AZ: University of Arizona Press, 1985. ix, 209p. Includes index. Bibliography, p. 185-206.

NAL Call no: QK211.N33

Other works by this author: *Enduring Seeds: Native American Agriculture and Wild Plant Conservation* (1989); *Ten Essential Reasons to Protect the Birds and the Bees: How an Impending Pollination Crisis Threatens Plants and the Food on Your Table,* with Mrill Ingram and Stephen L. Buchmann (1996); *Forgotten Pollinators,* with Stephen L. Buchmann (1996); *Coming Home to Eat: The Pleasures and Politics of Local Foods* (2002); *Conserving Migratory Pollinators and Nectar Corridors in Western North America,* editor (2004).

Annotation: Nabhan's books about plant and animal conservation are written with a personal touch that inspires personal action. This one looks at plants of the Sonoran Desert and their Native American uses. He has also published books about traditional food sources and on plant pollinating organisms. MVG

1986

Gever, John, Robert Kaufman, David Shole, Charles Vorosmarty and Carrying Capacity Inc.

Beyond Oil: The Threat to Food and Fuel in the Coming Decades

Cambridge MA: Ballinger, 1986. A project of Carrying Capacity, Inc. Includes index. Bibliography, p. 257-285. Other editions: 3rd edition, 1991.

NAL Call no: TJ163.25.U6B44

Annotation: This study synthesizes geological, social and economic statistics to prediect probable energy, economic and agricultural outcomes into the next century. Divergent opinions from experts are included; specific chapters deal with energy use in agriculture. MVG

Cited in: Harwood (1990)

1986

Phipps, Tim, Pierre R. Crosson and Kent A. Price, editors

Agriculture and the Environment

Washington DC: National Center for Food and Agricultural Policy, Resources for the Future, 1986. xvii, 298p. Includes bibliographies.

NAL Call no: HD1755.A44

Annotation: Based on papers and discussion at a Conference on Agriculture and the Environment sponsored by the National Center for Food and Agricultural Policy held in April 1986 at Resources for the Future, Washington, D.C. Papers focus on identifying the environmental problems of conventional agriculture and evaluating related programs and policy. MVG

Cited in: Harwood (1990)

1987

American Farmland Trust

Farming on the Fringe

New York: American Map Corp., 1987. 1 mapp. Relief shown by shading. Shows high market value farming counties. Concept and analysis for AFT, Margaret Stewart Maizel; design and production, Chessler and Associates. Includes text, local map insets and charts. Includes bibliographical references.

NAL Call no: ArU G3701.J15 1987

Annotation: Since 1980, American Farmland Trust has focused on the threats facing agricultural lands in the U.S.: unplanned, sprawling development; inadequate conservation programs; and a lack of options for farmers and ranchers who want to stay on their lands. This graphic presentation identifies valuable agricultural land near metropolitan areas and thus seriously threatened.

Comprehensive follow-up studies in 1994 and 1997 augment this information, using data from the U.S. Census and the U.S. Department of Agriculture. MVG. See also: Sorensen, 1997.

1987

California Certified Organic Farmers (CCOF)

California Certified Organic Farmers 1989 Certification Handbook

Santa Cruz CA: CCOF, 1989. vi, 21, 17p. Includes bibliographical references.

NAL Call no: S605.5.C3

Annotation: 1987 marked the first edition of the CCOF Certification Handbook and Materials List and the first Farm Inspection Manual, as well as the first series of Farm Inspector Trainings. This pioneering organization, formed by a group of grassroots activist farmers in 1973, is one of the oldest organic certifiers in North America. MVG

1987

Murphy, Bill

Greener Pastures on Your Side of the Fence: Better Farming with Voisin Grazing Management

Colchester VT: Arriba Publishing, 1987. xvi, 215p.

NAL Call no: SB199.M87

Annotation: Murphy put into American farmer-friendly terms and practices, the rational/rotational grazing techniques developed by Voisin in France. MVG. See also: Voisin, 1957.

1987

Whatley, Booker T. and New Farm, 1915-2005

Booker T. Whatley's Handbook on How to Make 100,000 Farming 25 Acres: With Special Plans for Prospering on 10 to 200 Acres

Emmaus PA: Regenerative Agriculture Association; distributed by Rodale Press, 1987. ix, 180p. Edited by George DeVault.

NAL Call no: S501.2.W47

Annotation: Whatley, a horticulture professor at Tuskegee University, presented practical, positive enterprise options for small farm operators including farm diversification, organic farming practices, farm value-added products and innovative, direct marketing schemes. These farming alternatives have grown and flourished, and currently provide important niche markets for small- and medium-sized farms. MVG

1988

Anon.

Nature's Ag School: The Thompson Farm

Emmaus PA: Regenerative Agriculture Association, 1988. 65p.

NAL Call no: S605.5.N37 1988

See also: On-farm Research Reports, by Dick Thompson, published annually in Boone IA, Thompson On-Farm Research, since 1993; and Oral History Interview with Dick Thompson, AFSIC (1991) (NAL Call #: Videocassette no.1008).

Annotation: Incorporating articles from New Farm magazine about regenerative agriculture, this volume introduced readers to Dick and Sharon Thompson's practical on-farm experiments, observations and educational opportunities. MVG

1988

Kirschenmann, Frederick

Switching to a Sustainable System: Strategies for Converting from Conventional/ Chemical to Sustainable/Organic Farming Systems

Windsor ND: Northern Plains Sustainable Agriculture Society, 1988. 18p.

NAL Call no: S494.5.S86K5

Other works by this author: Biotechnology and Sustainable Agriculture (1992); *On Becoming Lovers of the Soil* (1994). See also: Oral History Interview with Fred Kirschenmann, AFSIC (1990) (NAL Call #: Videocassette no.877).

Annotation: For many years, this information-packed book was the only practical reference for farmers, providing strategies and financial scenarios for conversion to more sustainable practices. Based at Kirschenmann Family Farms in North Dakota, Mr. Kirschenmann has been a sustainable agriculture pioneer and long-term advocate, serving with many regional and national organizations including the U.S. Department of Agriculture's National Organic Standards Board and the U.S. Department of Agriculture's North Central Region's Sustainable Agriculture Research and Education (SARE) program. MVG

1988

Madden, James Patrick, James A. DeShazer, Frederick R. Magdoff, Charles W. Laughlin and David E. Schlegel, editors

Low-input/Sustainable Agriculture Research and Education Projects for 1988

Washington DC: Low-Input Project, Cooperative State Research Service, U.S. Dept. of Agriculture, 1988. 28p.

NAL Call no: aS494.5 S86L68 1988

Other works by Madden: Research Agenda for the Transition to a Regenerative Food System, with Richard R. Harwood (1982); For All Generations: Making World Agriculture More Sustainable, editor, with Scott G. Chaplowe (1997); The Early Years of the LISA, SARE, and ACE Programs: Reflections of the Founding Director (1998) (full-text: SARE Western Region, http://wsare.usu.edu/about/index.cfm?sub=hist_sare) (accessed Apr. 23, 2007). See also: Oral History Interview with James Patrick Madden, AFSIC (1990) (NAL Call #: Videocassette no.1009).

Annotation: This report catalogs and describes the first projects funded under USDA's Low-input/Sustainable Agriculture (LISA) Research and Education program, initiated in 1985. This program became the Sustainable Agriculture Research and Education (SARE) program in 1988. Madden, an agricultural economist, was an early advocate of sustainable agriculture in the U.S. He was instrumental in creating and leading the LISA and SARE programs, and he co-authored the historic National Academy of Sciences book, Alternative Agriculture (National Academy Press, 1989). MVG

Cited in: Madden (1998)

1988

Savory, Allan, 1935-

Holistic Resource Management

Washington DC: Island Press, 1988. xxvi, 564p. Includes index. Bibliography, p. 513-539.

NAL Call no: HC59.S33

Other works by this author: *Holistic Resource Management Workbook,* with Sam Bingham (1990); *Holistic Management: A New Framework for Decision Making,* with Jody Butterfield (1999); *Holistic Management Handbook: Health Land, Healthy Profits,* with Jody Butterfield, Sam Bingham (2006).

Annotation: "In studying our ecosystem and the many creatures inhabiting it we cannot meaningfully isolate anything, let alone control the variables. The earth's atmosphere, its plant, animal and human inhabitants, its oceans, plains and forests, its ecological stability and its promise for humankind can only be grasped when they are viewed in their entirety. Isolate any part and neither what you have taken nor what you have left behind remains what it was when all was one." Now called "holistic management," Savory's model has been especially effective in supporting sustainable grazing and livestock systems. MVG

1988

Van En, Robyn

Basic Formula to Create Community Supported Agriculture

Great Barrington MA: R. Van En, 1992. 1 vol. Includes bibliographical references.

NAL Call no: HD9225.A2V35 1992

Other works by this author: Sharing the Harvest: A Guide to Community Supported Agriculture, with Elizabeth Henderson (1999).

Annotation: The first edition of this brief how-to publication appeared in 1988 and described a new, cooperative approach to marketing farm products as implemented at the author's Indian Line Farm in Massachusetts. This concept of "community supported agriculture," or CSA, had evolved from producer-consumer food alliances in Northern Europe, Japan and Chile of the 1970s. CSA, in a variety of applications, has since become a flourishing direct marketing option for small, organic farms in the United States and many other countries. MVG

1989

Coleman, Eliot, 1938-

The New Organic Grower: A Master's Manual of Tools and Techniques for the Home and Market Gardener

Chelsea VT: Chelsea Green, 1989. xv, 269p. Includes index. Bibliography, p. 225-235. Revised and expanded, 1995.

NAL Call no: SB324.3.C65

Other works by this author: *The New Organic Grower's Four Season Harvest: How-to Harvest Fresh Organic Vegetables from your Home Garden All Year Long* (1992); The *Winter-harvest Manual: Farming the Back Side of the Calendar: Commercial Greenhouse Production of Fresh Vegetables in Cold-winter Climates without Supplementary Heat* (1998); *Four-season Harvest: Organic Vegetables from your Home Garden All Year Long* (1998).

Annotation: Coleman's involvement with organic agriculture has been long-term. He is a farmer and researcher who has explored alternative production techniques and marketing practices; he is an educator and advisor who has written extensively, hosted educational television shows, directed the International Federation of Organic Agriculture Movements (IFOAM) and participated in publishing the 1980 USDA Report and Recommendations on Organic Farming (cited elsewhere in this bibliography). His practical guides, especially those on cold-climate production, have become mainstays for growers and students. MVG

Cited in: Kuepper, Gegner (2004)

1989

National Research Council Committee on the Role of Alternative Farming Methods in Modern Production Agriculture

Alternative Agriculture

Washington DC: National Academy Press, 1989. xiv, 448p. Includes bibliographies and index.

NAL Call no: S441.A46

Full-text: National Academy Press (NAP), http://books.nap.edu/catalog/1208.html#toc (accessed Jan. 1, 2007)

Annotation: This landmark study, compiled by the nation's foremost body of agricultural scientists, the National Academy of Sciences' Board on Agriculture, lent validity and immediacy to sustainable agriculture issues. MVG

Cited in: Kirschenmann (2004); Madden (1998)

1990

Edwards, Clive A., Rattan Lal, James Patrick Madden, Robert H. Miller and Gar House, editors

Sustainable Agricultural Systems

Ankeny IA: Soil and Water Conservation Society, 1990. xvi, 696p. 40 papers.

NAL Call no: S494.5 S86S86

Annotation: These papers come from the Proceedings of the International Conference on Sustainable Agricultural Systems held under the sponsorship of the United States Agency for International Development (USAID) at Ohio State University, Columbus, Ohio, in September 1988. They present a synthesis of 1980s sustainable agricultural science and philosophy in the words of dozens of researchers from around the world. MVG

1990

Gliessman, Stephen R.

Agroecology: Researching the Ecological Basis for Sustainable Agriculture

New York: Springer-Verlag, 1990. xiv, 380p. Includes bibliographical references and index. (Ecological Studies, 78)

NAL Call no: QH540.E288 v.78

Other works by this author: *Agroecology: Ecological Processes in Sustainable Agriculture*, editor (1998); *Agroecosystem Sustainability: Developing Practical Strategies* (2001); *Field and Laboratory Investigations in Agroecology* (2006); *Agroecology: The Ecology of Sustainable Food Systems* (rev. ed., 2007).

Annotation: An introduction to research approaches and case studies in the emerging interdisciplinary field of agroecology. MVG

1990

Lampkin, Nicolas

Organic Farming

Ipswich, UK; Alexandria Bay NY: Farming Press; distributed in North America by Diamond Farm Enterprises, 1990. xiii, 701p. Includes bibliographical references (p. 655-656) and index. Other editions: Rev. 1992; reprinted with amendments, 2002.

NAL Call no: S605.5.L35 1990

Other works by this author: *The Soil: Assessment, Analysis and Utilisation in Organic Agriculture*, editor, with L. Woodward (1991); *The Economics of Organic Farming: An International Perspective*, editor, with S. Padel (1994); *1994 Organic Farm Management Handbook*, editor, with Mark Measures (1st ed., 1994); *Impact of EC Regulation 2078/92 on the Development of Organic Farming in the European Union* (1996); *The Policy and Regulatory Environment for Organic Farming in Europe: Country Reports*, with Carolyn Foster, Susanne Padel (1999); *European Organic Production Statistics, 1993-1996*, with Carolyn Foster (1999).

Annotation: Dr. Lampkin's 1990 work (updated in 2002) remains one of the most comprehensive texts on organics, addressing all aspects of organic crop and livestock production. He is currently director of the Organic Centre Wales where he has long been active in researching and writing about the: "Economics of organic farming and specifically conversion to organic methods. Role of organic farming in agricultural policy, in particular impacts of CAP on organic farming and contribution of organic farming to agri-environmental policy. Dissemination of information to key user groups." Institute of Rural Studies Staff Webpage (http://www.irs.aber.ac.uk/research/expertise/nhl.shtml (Accessed May 7, 2007). MVG

1990

United States Congress

The Organic Foods Production Act (OFPA)

Washington DC: Government Printing Office, 1990. CFR Title 7, Chapter I, Agricultural Marketing Service (Standards, Inspections, Marketing Practices), Department of Agriculture, Part 205 (as authorized under the Organic Foods Production Act of 1990, as amended).

NAL Call no: n.a.

Full-text: USDA National Organic Program (NOP), http://www.ams.usda.gov/nop/NOP/standards.html (accessed Jan. 1, 2007)

Annotation: The Organic Foods Production Act (OFPA) of 1990 mandated the U.S. Department of Agriculture (USDA) to develop and maintain national standards for organically produced agricultural products to assure consumers that agricultural products marketed as organic meet consistent, uniform standards. The OFPA and the National Organic Program (NOP) regulations require that agricultural products labeled as organic originate from farms or handling operations certified by a State or private entity that has been accredited by USDA. NOP standards were fully implemented in October 2002. MVG

Cited in: Heckman (2006); Kirschenmann (2004)

1990

United States Congress

Food, Agriculture, Conservation and Trade Act of 1990 (FACTA)

Washington DC: Government Printing Office, 1990. 101st Congress, 2nd Session. S. 2334. "To expand the United States Department of Agriculture's low input sustainable agriculture research and education programs, and to provide biotechnology risk assessment research, and for other purposes." SARE National Project Database is online at http://www.sare.org/ (accessed Feb. 1, 2007).

NAL Call no: KF1692.A31 1990

Full-text: Sustainable Agriculture Research and Education (SARE), Western Region, http://wsare.usu.edu/about/index.cfm?sub=FACTA (accessed Jan. 1, 2007)

Annotation: Subtitle B of Title XVI of this bill authorized the USDA Sustainable Agriculture Research and Education (SARE) program. It expanded 1985 legislation that had implemented the USDA Low-Input Sustainable Agriculture (LISA) program. Thousands of research and extension projects have been funded since LISA and SARE became operational in 1988. MVG

Cited in: Heckman (2006); Kirschenmann (2004)

1990

United States General Accounting Office

Alternative Agriculture: Federal Incentives and Farmers' Opinions: Report to Congressional Requesters

Washington DC: The Office, 1990. 95p. (GAO/PEMD-90-12)

NAL Call no: S494.5.A65U54

Annotation: This government report documented, through surveys and analysis, farmer and citizen concerns about chemical pesticide use in food production. It helped support U.S. legislation aimed at sustainable farming policy and research. MVG

Cited in: Madden (1998)

1992

Goreham, Gary, David L. Watt and Roy M. Jacobsen

The Socioeconomics of Sustainable Agriculture: An Annotated Bibliography

New York: Garland, 1992. xix, 334p. (Garland Reference Library of the Humanities, 1332)

NAL Call no: Z5074 E3G69 1992

Annotation: This compilation lists and describes books and book chapters, journal articles, conference proceedings, government documents and research reports from the 1970s through 1992. Indexed by author and keyword. MVG

1992

Soule, Judith D. and Jon K. Piper

Farming in Nature's Image: An Ecological Approach to Agriculture

Washington DC: Island Press, 1992. xix, 286p. Foreword by Wes Jackson. Includes bibliographical references (p. 231-278) and index.

NAL Call no: S441.S757

Annotation: Based in 1992 at the Land Institute in Kansas, the authors present a "scientifically backed plea for the radically new form of agriculture first laid out in Wes Jackson's New Roots for Agriculture. ...the economic problems of farmers and rural communities and the environmental damage resulting from modern agricultural practices, have roots in the industrialization of agriculture." MVG

Cited in: Esbjornson (1992); Heckman (2006)

1993

Ikerd, John E.

Assessing the Health of Agroecosystems: A Socieoeconomic Perspective

Columbia: University of Missouri, 1993. "This paper was presented at the first International Ecosystem Health and Medicine Symposium as part of a panel organized by Kathryn Freemark and Jack Waide and sponsored by the Canadian Wildlife Service of Environment, Canada, Ottawa."

NAL Call no: n.a.

Full-text: Recent Papers, http://web.missouri.edu/~ikerdj/papers/Otta-ssp.htm (accessed Jan. 1, 2007)

Other works by this author: *Sustainable Capitalism: A Matter of Common Sense* (2005).

Annotation: Dr. Ikerd's understanding of sustainable agriculture grew out of his observations as an agricultural economist. Now retired from the University of Missouri, his writings discuss sustainability in terms of small farms, organic farming, rural communities, local food systems and more. He helped implement the 1,000 Ways to Sustainable Farming project funded by USDA's Sustainable Agriculture Research and Education Program (SARE) that culminated in the book, The New American Farmer: Profiles of Agricultural Innovation (2001) (full-text: Sustainable Agriculture Network, http://www.sare.org/publications/naf.htm) (accessed Apr. 23, 2007). MVG

1994

Bonanno, Alessandro, editor

From Columbus to ConAgra: The Globalization of Agriculture and Food

Lawrence KS: University Press of Kansas, 1994. viii, 294p. Includes bibliographical references and index. (Rural America)

NAL Call no: HD9000.5.F76 1994

Annotation: Food systems and attached sustainability issues are increasingly tied to global trade and regulation. This collection of essays includes works by William D. Heffernan and Douglas H. Constance ("Transnational

Corporations and the Globalization of the Food System"); Lawrence Busch ("The State of Agricultural Science and the Agricultural Science of the State"); Lourdes Gouveia ("Global Strategies and Local Linkages: The Case of the U.S. Meatpacking Industry"); and William H. Friedland ("The New Globalization: The Case of Fresh Produce"). MVG

1994

National Research Council Committee on Rangeland Classification

Rangeland Health: New Methods to Classify, Inventory, and Monitor Rangelands

Washington DC: National Academy Press, 1994. xvi, 180p.

NAL Call no: SF85.3.R36 1994

Full-text: National Academy Press, http://www.nap.edu/books/0309048796/html/index.html (accessed Jan. 1, 2007)

Annotation: Understanding and managing sustainable use of U.S. rangelands has been hampered by a lack of uniform methods for inventorying, classifying and monitoring rangelands. The NRC Committee on Rangeland Classification Systems was tasked with examining the scientific basis of methods used by three U.S. agencies—the Soil Conservation Service, Bureau of Land Management and U.S. Forest Service. Since its publication, interagency cooperation has produced standard measurement tools based on multiple ecological indicators to evaluate rangeland health and sustainability. MVG

1995

Benson, Laura Lee, Robert Zirkel and Kickapoo Organic Resource Network

Organic Dairy Farming

Gays Mills WI: Orang-utan Press, 1995. 87p. Includes bibliographical references. Illustrations by Pam Taliaferro.

NAL Call no: SF239.B46 1995

Annotation: This slim volume was one of the first to provide farmers with practical information about organic livestock husbandry and organic milk production. MVG

1995

Pretty, Jules N.

Regenerating Agriculture: Policies and Practice for Sustainability and Self-reliance

Washington DC: Joseph Henry Press, 1995. 336p. Includes bibliographical references and index.

NAL Call no: S464.5 S86P74 1995

Other works by this author: *Unwelcome Harvest: Agriculture and Pollution*, with Gordon R. Conway (1991); *The Living Land: Agriculture, Food and Community Regeneration in Rural Europe* (1998); *Agri-culture: Reconnecting People, Land and Nature* (2002); *The Earthscan Reader in Sustainable Agriculture*, editor (2005)

Annotation: Pretty has documented case studies from around the developing world, illustrating the techniques and impacts of successful sustainable farming practices in Brazil, Burkina Faso, Honduras, India, Indonesia, Kenya, Lesotho, Mali, Mexico, Peru, Philippines and Sri Lanka. MVG

1996

Smit, Jac, Annu Ratta, Joe Nasr and United Nations Development Programme

Urban Agriculture: Food, Jobs and Sustainable Cities

New York: United Nations Development Programme, 1996. xxi, 300p. Includes bibliographical references.

NAL Call no: S494.5.U72U73 1996

Annotation: This book portrays crop and livestock production in cities and suburbs as a modern economic activity with significance for sustainable food systems - ecological production, food security, stable family incomes and a livable urban environment. It is based on case studies from Asia, Africa and Latin America. MVG

1997

Abdul-Baki, Aref A. and John R. Teasdale

Sustainable Production of Fresh-Market Tomatoes and Other Summer Vegetables with Organic Mulches

Beltsville MD: U.S. Department of Agriculture, Agricultural Research Service, 1997. 23p. Includes bibliographical references. (Farmers' Bulletin, 2279)

NAL Call no: 1 Ag84F no.2279

Full-text: USDA, ARS, http://www.ars.usda.gov/is/np/SustainableTomato.pdf (accessed Jan. 1, 2007)

Annotation: Tomatoes grown with a hairy vetch cover crop/mulch out-produced conventionally-grown tomatoes and required significantly fewer inputs. This small book summarizes the research, done by USDA scientists at the Agricultural Research Service in Maryland and provides growers with practical information on using the system on their own farms. MVG
Cited in: Kuepper, Gegner (2004)

1997

Lipson, Mark

Searching for the "O-word": Analyzing the USDA Current Research Information System for Pertinence to Organic Farming

Santa Cruz CA: Organic Farming Research Foundation (OFRF), 1997. 83p. Includes bibliographical references, p. 82.

NAL Call no: S605.5.L56 1997

Other works from the Foundation: *State of the States: Organic Systems Research at Land Grant Institutions, 2001-2003* (2003); *The Fourth National Organic Farmers' Survey: Sustaining Organic Farms in a Changing Organic Marketplace* (2004) (both in full-text: OFRF, http://ofrf.org/publications/publications.html) (accessed Feb. 1, 2007).

Annotation: Lipson's documented catalog and analysis of organic-pertinent research being done (and not being done) with USDA funds, received the attention of senior research staff and administrators in the Agriculture Department. Spurred by this book and by the growing commercial success of organically-produced commodities, Congress and USDA began to implement more organic systems research and marketing support. MVG

1997

Sorensen, A. Ann, Richard P. Greene and Karen Russ

Farming on the Edge

Washington DC; DeKalb IL: American Farmland Trust; Center for Agriculture in the Environment, 1997. 66p. Features maps, tables and a glossary. Foldout map included.

NAL Call no: HD256.S6 1997

Full-text: American Farmland Trust (partial report), http://www.farmland.org/resources/fote/default.asp (accessed Jan. 1, 2007)

Annotation: This report utilizes many maps and tables to identify agricultural lands near metropolitan areas and how they are threatened, as of 1997. Updates data and reports from 1987 and 1994. MVG. See also: American Farmland Trust, 1987.

1998

United States Department of Agriculture National Commission on Small Farms

A Time to Act: A Report of the USDA National Commission on Small Farms

Washington DC: The Commission, 1998. 121p. Includes bibliographical references. (Miscellaneous Publication, 1545)

NAL Call no: 1 Ag84M no.1545

Full-text: USDA, http://www.csrees.usda.gov/nea/ag_systems/in_focus/smallfarms_if_time.html (accessed Jan. 1, 2007)

Annotation: The plight of small and family farms became so grave during the 1980s and 1990s that a 30-member National Commission on Small Farms was appointed by the Secretary of Agriculture in July 1997 to examine the status of small farms in the United States and to determine a course of action for USDA to recognize, respect and respond to their needs. The report established eight policy goals. MVG

1999

Dobson, Andrew P., editor

Fairness and Futurity: Essays on Environmental Sustainability and Social Justice

Oxford: Oxford University Press, 1999. 344p.

NAL Call no: n.a.

Annotation: This impressive collection of essays comes from collaborative work done by environmental and political theorists, researchers and public policy makers. Contributors have addressed difficult questions about the concepts and goals of sustainability and justice, many of which related directly to agricultural production: "If future generations are owed justice, what should we bequeath them? Is 'sustainability' an appropriate medium for environmentalists to express their demands? Is environmental protection compatible with intragenerational justice? Is environmental sustainability a luxury when social peace has broken down?" Publisher's Web page (http://www.oup.com/us/catalog/general/subject/Politics/PoliticalTheory/PoliticalPhilosophy/?view=usa&ci=9780198294894) (accessed Apr. 23, 2007). MVG

2000

Magdoff, Fred, John Bellamy Foster and Frederick H. Buttel, editors

Hungry For Profit: The Agribusiness Threat to Farmers, Food, and the Environment

New York: Monthly Review Press, 2000. 248p.

NAL Call no: HD9000.5 .H86 2000

Annotation: This collection of provocative essays covers many aspects of agriculture and agribusiness. Essay titles include: "Liebig, Marx, and the Depletion of Soil Fertility: Relevance for Today's Agriculture" (John Bellamy Foster and Fred Magdoff); "Concentration of Ownership and Control in Agriculture" (William D. Heffernan); "Ecological Impacts of Industrial Agriculture and the Possibilities for Truly Sustainable Farming" (Miguel A. Altieri); "New Agricultural Biotechnologies: The Struggle for Democratic Choice" (Gerard Middendorf et al.); "Organizing U.S. Farm Workers: A Continuous Struggle" (Linda C. Majka and Theo J. Majka); "Rebuilding Local Food Systems from the Grassroots Up" (Elizabeth Henderson); "Want Amid Plenty: From Hunger to Inequality" (Janet Poppendieck); "Cuba: A Successful Case Study of Sustainable Agriculture" (Peter M. Rosset); "The Importance of Land Reform in the Reconstruction of China" (Wiliam Hinton). MVG

2001

Furuno, Takao, 1950-

The Power of Duck: Integrated Rice and Duck Farming

Tasmania, Australia: Tagari Publications, 2001.

NAL Call no: n.a.

Other works by this author: *Aigamo Banzai: Aigamo Suito Doji Saku No Jissai* [transliterated title, in Japanese, translated title: Cheers for Aigamo Ducks] (1992).

Annotation: Japanese rice farmer Furuno has developed a system that eliminates external fertilizer and pesticide inputs by carefully integrating the production of rice, ducks and fish. Profits from the resulting increased rice yields, fish, duck meat and eggs have persuaded thousands of farmers in Asia to switch to Furuno's system. MVG

Cited in: Kirschenmann (2004)

2001

Horne, James E. and Maura McDermott

The Next Green Revolution: Essential Steps to a Healthy, Sustainable Agriculture

New York: Food Products Press, 2001. xix, 312p. Includes bibliographical references, p. 285-297, and index.

NAL Call no: S441 .H67 2001

Other works by James Horne: *Laboratory Experiments in Sustainable Agriculture Wheat for Pasture: Mineral Nutrient Uptake*, with Alexandre E. Kalevitch, Mariia V. Filimonova (1994); *Germination of Five Wheat Varieties under Various Soil Conditions: Implications for Sustainable Agriculture*, with Alexandre E. Kalevitch, Mariia V. Filimonova (1994); *Seeds of Change: Food and Agriculture Policy for Oklahoma's Future*, with Anita K. Poole (2003). Other works by Maura McDermott: *Hoeing the Row Out: More than Three Decades of History* (1996); *Future Farms 2002: A Supermarket of Ideas: Conference Proceedings, November 15 and 16, 2002* (2003).

Annotation: An educator and economist, much of Horne's career has been at the Kerr Center for Sustainable Agriculture in Oklahoma where he is currently President and Chief Executive Officer. In this book, he and McDermott share what has been "learned as the Kerr Center experimented with new 'sustainable' approaches to old problems on the Center's ranch/farm and ... experiences working with the USDA's Sustainable Agriculture Research and Education

Program. It gives practical suggestions for increasing profits and reducing risks while regenerating the soil, protecting the environment and being a good neighbor." Kerr Center Web site (http://www.kerrcenter.com/staff/horne.html) (accessed 4/12/2007). MVG

2001

Petrini, Carlo, Benjamin Watson and Slow Food Movement, editors

Slow Food: Collected Thoughts on Taste, Tradition, and the Honest Pleasures of Food

White River Junction VT: Chelsea Green, 2001. xv, 287p. Foreword by Deborah Madison. Includes bibliographical references.

NAL Call no: TX631 .S58 2001

Other works by Petrini: *Slow Food: The Case for Taste* (2004); *Slow Food Revolution: A New Culture for Eating and Living*, with Gigi Padovani (2006); *Slow Food Nation: A Blueprint for Changing the Way We Eat* (2007).

Annotation: The "slow food" movement, founded by Petrini in 1986, emphasizes regional food and home cooking that uses sustainably grown ingredients grown and harvested by fairly treated workers. The organization, which originated in Italy, has grown worldwide and now (2007) has more than 80,000 members. MVG

2001

Salatin, Joel. F.

Family Friendly Farming: A Multi-generational Home-based Business Testament

Swoope VA: Polyface, 2001. xvii, 402p. Includes bibliographical references, p. 390-394, and index.

NAL Call no: HD1476.U6 S24 2001

Other works by this author: *Pastured Poultry Manual: The Polyface Model* (1991); *Pastured Poultry Profits* (1993); *Salad Bar Beef* (1995); *You Can Farm: The Entrepreneur's Guide to Start and Succeed in a Farm Enterprise* (1998).

Annotation: Salatin, a Virginia farmer, published his first book in 1991 (cited above); it described his successful pastured poultry operation. Since then, he has built on and refined his farming enterprises, lectured widely about his successes and written several how-to books. His combination of sustainable farming techniques and profitability has inspired and influenced farmers across the U.S. MVG

2002

Norberg-Hodge, Helena, Todd Merrifield and Steven Gorelick

Bringing the Food Economy Home: Local Alternatives to Global Agribusiness

London; Halifax NS; Bloomfield CT: Zed Books; Fernwood; Kumarian Press, 2002. vi, 150p. Published in association with the International Society for Ecology and Culture (ISEC). Includes bibliographical references, p. 127-142, and index.

NAL Call no: HD9000.5 .N596 2002

Other works by Norberg-Hodge: *Ancient Futures: Learning from Ladakh* (1991); *From the Ground Up: Rethinking Industrial Agriculture,* with Peter Goering, John Page (1993; revised edition, 2001).

Annotation: Since 2000, several books have focused on the risks and remedies related to sustainable food systems and market globalization. This one covers a wide range of issues with particular reference to the U.S. and U.K. and points to locally-based alternatives to global consumer culture. MVG

2005

Diamond, Jared

Collapse: How Societies Choose to Fail or Succeed

New York: Viking Press, 2005. 575p. "Further Readings" appendix, p. 529-560.

NAL Call no: n.a.

Other works by this author: *Guns, Germs and Steel: The Fates of Human Societies* (1997).

Annotation: Diamond's discussion of ecocide, "people inadvertently destroying the environmental resources on which their societies depended," emphasizes food system resources. He looks at problematic ancient and contemporary societies, discusses similarities and differences and provides a chapter called, "Reasons for Hope." "Our television documentaries and books show us in graphic detail why the Easter Islanders, Classic Maya and other past societies collapsed. Thus, we have the opportunity to learn from mistakes of distant peoples and past peoples. That's an opportunity that no past society enjoyed to such a degree." MVG

2005

Hatfield, Jerry L., editor

The Farmer's Decision: Balancing Economic Agriculture Production with Environmental Quality

Ankeny IA: Soil and Water Conservation Society, 2005. xviii, 251p. Essays based on presentations at a workshop in Honolulu, Hawaii, held November 9-12, 2004. Includes bibliographical references and index.

NAL Call no: S589.75 .F36 2005

Other works by this author: Sustainable Agriculture Systems, editor, with D.L. Karlen (1984); *Precision Agriculture and Environmental Quality: Challenges for Research and Education* (2000); *Nitrogen in the Environment: Sources, Problems, and Management*, editor, with R.F. Follett (2001).

Annotation: This book examines the decision making processes involved in moving scientists, policymakers, planners, producers, and the consumers of food and fiber to a more sustainable agriculture and healthier environment. Thirteen essays discuss how farmers make decisions and adopt innovation, as well as the environmental, economic, social and technological issues driving those decisions, e.g. water and nutrient management challenges, precision farming, etc. MVG

2006

Francis, Charles A., Raymond P. Poincelot and George W. Bird, editors

Developing and Extending Sustainable Agriculture: A New Social Contract

New York : Haworth Food and Agricultural Products Press, 2006. xxii, 367p. Includes bibliographical references and index. (Sustainable Food, Fiber, and Forestry Systems)

NAL Call no: S494.5.S86 D48 2006

Full-text: Haworth Press (selected text only), http://www.haworthpress.com/store/SampleText/5709.pdf (accessed Jan. 1, 2007)

Other works by Charles Francis: *Enough Food: Achieving Food Security through Regenerative Agriculture,* with Richard R. Harwood (1985); *Multiple Cropping Systems* (1986); *Internal Resources for Sustainable Agriculture* (1988); *Sustainable Agriculture in the Midwest*: North Central Regional Conference, editor (1988); *Sustainable Agriculture in Temperate Zones,* editor (1990) See also: Oral History Interview with Charles A. Francis, AFSIC (1990) (NAL Call #: Videocassette no.876). Other works by Raymond Poincelot: The Biochemistry and Methodology of Composting (1972, rev. 1975); Toward a More Sustainable Agriculture (1986); Sustainable Horticulture: Today and Tomorrow (2004). Other works by George Bird: Sustainable Agriculture: Current State and Future Trajectory (1988); Integrated Pest Management: Its Role in the Evolution of Sustainable Agriculture (1991).

Annotation: This book explores integrating agricultural research and outreach with the goal of moving U.S. farming and ranching to practical, profitable, sustainable systems. Contributors, including the editors, are leaders in the sustainable agriculture field. They present new thoughts on soil management, managed grazing, whole-farm planning, economic issues, integrated pest management and community/rural development. Special attention is given to the status and potential of research, education and outreach efforts. MVG

2006

Kristiansen, Paul, Acram Taji and John Reganold, editors

Organic Agriculture: A Global Perspective

Collingwood, Vic.: CSIRO Publishing, 2006. xxxi, 449p. Includes bibliographical references and index.

NAL Call no: S605.5 .O62 2006

Other works by John Reganold: *Natural Resource Conservation: Management for a Sustainable Future,* with Oliver S. Owen, Daniel D. Chiras (7th ed., 1998).

Annotation: "For more than two decades, research into organic methods by mainstream scientists has generated a large body of information that can now be integrated and used for assessing the actual impacts of organic farming in a wide range of disciplines. The knowledge of selected international experts has been combined in one volume, providing a comprehensive review of organic farming globally... At the intersection of research, education, and practice, the contributors look at the organic agricultural movement's successes and limitations." CSIRO Web page (http://www.publish.csiro.au/pid/5325.htm) (accessed 4/23/2007). MVG

2006

Pollan, Michael

The Omnivore's Dilemma: A Natural History of Four Meals

New York: Penguin Press, 2006. 450p. Includes bibliographical references, p. 417-435, and index.

NAL Call no: GT2850 .P65 2006

Other works by this author: *Botany of Desire: A Plant's Eye View of the World* (2001).

Annotation: Pollan writes eloquently about where our food comes from and what its production (and our choice to eat it) entails ecologically, economically and nutritionally. He follows and analyzes four food chains—industrial food, organic or alternative food and food we forage ourselves—from the soil to the final meal. MVG

2007

Newton, Paul C. D., Carran R. Andrew, Grant R. Edward and Pascal A. Niklaus, editors

Agroecosystems in a Changing Climate

Boca Raton FL: CRC/Taylor and Francis, 2007. 364p. Includes bibliographical references and index. (Advances in Agroecology)

NAL Call no: S589.7 .A374 2007

Annotation: No topic could be more pertinent to sustainable agriculture than global climate change. This work combines information from both ecological and agricultural perspectives. Essays discuss how atmospheric and climate changes will effect biogeochemical cycles, nutrient and water supply; symbiotic nitrogen fixation; belowground food webs; herbivory and nutrient cycling; plant population dynamics and species composition; fungi; trophic interactions;

weed, pest and disease problems for plants; and plant breeding. There is also an essay on "Distinguishing between acclimation and adaptation." MVG

2007

Warner, Keith Douglass
Agroecology in Action: Extending Alternative Agriculture through Social Networks
Cambridge MA: MIT, 2007. xiv, 273p. Includes bibliographical references (p. 257-270) and index. Forward by Frederick L. Kirschenmann. (Food, Health, and the Environment)
NAL Call no: S441 .W28 2007
Other works by this author: *Promise or Threat? Responding to the Challenge of Agricultural Biotechnology* (2000) (full-text: National Catholic Rural Life Conference, http://www.ncrlc.com/promise_threat.html) (accessed Apr. 23, 2007).
Annotation: Partnerships are keys to establishing a sustainable agriculture. "This book describes how some farmers, scientists, agricultural organizations, and public agencies have developed innovative, ecologically informed techniques and new models of social learning to reduce reliance on agrochemicals, and thus put agroecology into action. It describes the technical scope and geographic range of ecologically informed strategies and practices in American agriculture, analyzing in detail a set of specific agroecological partnerships in California." Introduction. MVG

NOTES

1. The Conservation of Natural Resources. Washington DC: U.S. Dept. of Agriculture, 1908. (Farmers' Bulletin, 327) NAL Call no.: 1 Ag84F no.327
2. Food, Agriculture, Conservation, and Trade Act of 1990 (FACTA), Public Law 101-624, Title XVI, Subtitle A, Section 1603 Government Printing Office, Washington, DC, 1990. NAL Call # KF1692.A31 1990
3. The New American Farmer: Profiles of Agricultural Innovation, 2nd edition. Sustainable Agriculture Research and Education (SARE), 2006.
4. Information about the National Organic Program is available at http://www.ams.usda.gov/nop/indexNet.htm

5. For a lengthy discussion and list of references about the definition of sustainable agriculture and related terms, see the AFSIC publication, Sustainable Agriculture: Definitions and Terms, by Mary V. Gold, 1999 (http://afsic.nal.usda.gov/sustainable-agriculture-definitions-and-terms-1).
6. "In popular literature, sustainable agriculture generally is presented as a new phenomenon. Wes Jackson is credited with the first publication of the expression in his New Roots for Agriculture (1980), and the term didn't emerge in popular usage until the late 1980s." ("A Brief History of Sustainable Agriculture," by Fred Kirschenmann, in The Networker, vol. 9, no. 2, March 2004)
7. Collapse: How Societies Choose to Fail or Succeed. New York: Viking Press, 2005.

CHAPTER 12

Twenty-First-Century Organic and Sustainable Farming: A Brief Annotated Bibliography

Kim Etingoff

Alkon, Alison Hope and Julian Agyeman
Cultivating Food Justice: Race, Class, and Sustainability, MIT Press, 2011
Annotation: Popularized by such best-selling authors as Michael Pollan, Barbara Kingsolver, and Eric Schlosser, a growing food movement urges us to support sustainable agriculture by eating fresh food produced on local family farms. But many low-income neighborhoods and communities of color have been systematically deprived of access to healthy and sustainable food. These communities have been actively prevented from producing their own food and often live in "food deserts" where fast food is more common than fresh food. Cultivating Food Justice describes their efforts to envision and create environmentally sustainable and socially just alternatives to the food system. Bringing together insights from studies of environmental justice, sustainable agriculture, critical race theory, and food studies, Cultivating Food Justice highlights the ways race and class inequalities permeate the food system, from production to distribution to consumption. The studies offered in the book explore a range of important issues, including agricultural and land use policies that systematically disadvantage Native American, African American, Latino/a, and Asian American farmers and farmworkers; access problems in both urban and rural areas; efforts to create sustainable local food systems in low-income communities of color; and future directions for the food justice movement. These diverse

accounts of the relationships among food, environmentalism, justice, race, and identity will help guide efforts to achieve a just and sustainable agriculture.

Dawling, Pam
Sustainable Market Farming: Intensive Vegetable Production on a Few Acres, New Society Publishers, 2013
Annotation: This is the complete year-round guide for the small-scale market grower.

Gliessman, Stephen R. and Martha Rosemeyer
The Conversion to Sustainable Agriculture: Principles, Processes, and Practices, CRC Press, 2009
Annotation: With all of the environmental and social problems confronting our food systems today, it is apparent that none of the strategies we have relied on in the past—higher-yielding varieties, increased irrigation, inorganic fertilizers, pest damage reduction—can be counted on to come to the rescue. In fact, these solutions are now part of the problem. It is becoming quite clear that the only way to keep the food crisis from escalating is to promote the conversion processes that will move agriculture to sustainability.

Under the editorial guidance of agroecology experts Martha Rosemeyer and the internationally renowned Dr. Stephen R. Gliessman, The Conversion to Sustainable Agriculture: Principles, Processes, and Practices establishes a framework for how this conversion can be accomplished and presents case studies from around the world that illustrate how the process is already underway. The book provides a four-stage transition process for achieving sustainability and an in-depth analysis of the global efforts to make farms more energy-efficient and environmentally friendly.

An international team of chapter contributors explores ways to lessen dependency on fossil fuels and pesticides, and examines each step in the conversion process. They also describe the process of monitoring change toward sustainable agriculture while integrating social and economic analysis within scientific practices. Serving as both a core textbook for students and a comprehensive reference for agricultural practitioners, this volume is a valuable resource for the change that is needed in our food system now and in the future.

Guthman, Julie
Agrarian Dreams: The Paradox of Organic Farming in California, University of California Press, 2014

Annotation: In this groundbreaking study of organic farming, Julie Guthman challenges accepted wisdom about organic food and agriculture in the Golden State. Many continue to believe that small-scale organic farming is the answer to our environmental and health problems, but Guthman refutes popular portrayals that pit "small organic" against "big organic" and offers an alternative analysis that underscores the limits of an organic label as a pathway to transforming agriculture.

This second edition includes a thorough investigation of the federal organic program, a discussion of how the certification arena has continued to grow and change since its implementation, and an up-to-date guide to the structure of the organic farming sector. Agrarian Dreams delivers an indispensable examination of organic farming in California and will appeal to readers in a variety of areas, including food studies, agriculture, environmental studies, anthropology, sociology, geography, and history.

Halberg, Niels
Organic Agriculture for Sustainable Livelihoods, Routledge, 2013.
Annotation: This book provides a timely analysis and assessment of the potential of organic agriculture (OA) for rural development and the improvement of livelihoods. It focuses on smallholders in developing countries and in countries of economic transition, but there is also coverage of and comparisons with developed countries. It covers market-oriented approaches and challenges for OA as part of high value chains and as an agro-ecologically based development for improving food security. It demonstrates the often-unrecognized roles that organic farming can play in climate change, food security and sovereignty.

Ikerd, John E.
Crisis and Opportunity: Sustainability in American Agriculture, University of Nebraska Press, 2008
Annotation: With the decline of family farms and rural communities and the rise of corporate farming and the resulting environmental degradation, American agriculture is in crisis. But this crisis offers the opportunity to rethink agriculture in sustainable terms. Here one of the most eloquent and influential proponents of sustainable agriculture explains what this means. These engaging essays describe what sustainable agriculture is, why it began, and how it can succeed. Together they constitute a clear and compelling vision for rebalancing the ecological, economic, and social dimensions of agriculture to meet the needs of the present without compromising the future. John E. Ikerd outlines

the consequences of agricultural industrialization, then details the methods that can restore economic viability, ecological soundness, and social responsibility to our agricultural system and thus ensure sustainable agriculture as the foundation of a sustainable food system and a sustainable society.

Joffe, Daron
Citizen Farmers: The Biodynamic Way to Grow Healthy Food, Build Thriving Communities, and Give Back to the Earth, Abrams, 2014
Annotation: Biodynamic farming, with its focus on ecological sustainability, has emerged as the gold standard in the organic gardening movement. Daron Joffe (known as Farmer D) has made it his mission to empower, educate, and inspire people to become conscientious consumers, citizens, and stewards of the land. In this engaging call to action, Farmer D teaches us to not only create sustainable gardens but also to develop a more holistic, community-minded approach to how our food is grown and how we live our lives in balance with nature. Illustrated with photographs of gardens designed by Farmer D as well as line drawings, the book is an indispensable resource packed with advice on establishing a biodynamic garden, composting, soil composition and replenishment, controlling pests and disease, cooperative gardening practices, and even creating delicious meals.

Kingsolver, Barbara
Animal, Vegetable, Miracle: Our Year of Seasonal Eating, Faber & Faber, 2010
Annotation: "We wanted to live in a place that could feed us: where rain falls, crops grow, and drinking water bubbles up right out of the ground." Barbara Kingsolver opens her home to us, as she and her family attempt a year of eating only local food, much of it from their own garden. Inspired by the flavors and culinary arts of a local food culture, they explore many a farmers market and diversified organic farms at home and across the country. With characteristic warmth, Kingsolver shows us how to put food back at the centre of the political and family agenda. Animal, Vegetable, Miracle is part memoir, part journalistic investigation, and is full of original recipes that celebrate healthy eating, sustainability and the pleasures of good food.

Lengnick, Laura
Resilient Agriculture: Cultivating Food Systems for a Changing Climate, New Society Publishers, 2015

Annotation: Offers guidance on creating agile, resilient foodsheds to feed a world where climate change is an inescapable reality.

Lichtfouse, Eric
Organic Fertilisation, Soil Quality and Human Health, Springer Science & Business Media, 2012
Annotation: Sustainable agriculture is a rapidly growing field aiming at producing food and energy in a sustainable way for our children. This discipline addresses current issues such as climate change, increasing food and fuel prices, starvation, obesity, water pollution, soil erosion, fertility loss, pest control and biodiversity depletion. Novel solutions are proposed based on integrated knowledge from agronomy, soil science, molecular biology, chemistry, toxicology, ecology, economy, philosophy and social sciences. As actual society issues are now intertwined, sustainable agriculture will bring solutions to build a safer world. This book series analyzes current agricultural issues and proposes alternative solutions, consequently helping all scientists, decision-makers, professors, farmers and politicians wishing to build safe agriculture, energy and food systems for future generations.

Wright, Julia
Sustainable Agriculture and Food Security in an Era of Oil Scarcity: Lessons from Cuba, Earthscan, 2012
Annotation: When other nations are forced to rethink their agricultural and food security strategies in light of the post-peak oil debate, they only have one living example to draw from: that of Cuba in the 1990s. Based on the first and—up till now—only systematic and empirical study to come out of Cuba on this topic, this book examines how the nation successfully headed off its own food crisis after the dissolution of the Soviet Bloc in the early 1990s.

The author identifies the policies and practices required for such an achievement under conditions of petroleum-scarcity and in doing so, challenges the mainstream globalized and privatized food systems and food security strategies being driven through in both industrialized and more vulnerable developing regions. Paradoxically, the book dispels the myth that Cuba turned to organic farming nationwide, a myth founded on the success of Cuba's urban organic production systems which visitors to the country are most commonly exposed to. In rural regions, where the author had unique access, industrialized high-input and integrated agriculture is aspired to for the majority of domestic production, despite the ongoing fluctuations in availability of agrochemicals and fuel.

By identifying the challenges faced by Cuban institutions and individuals in de-industrializing their food and farming systems, this book provides crucial learning material for the current fledgling attempts at developing energy descent plans and at mainstreaming more organic food systems in industrialized nations. It also informs international policy on sustainable agriculture and food security for less-industrialized countries.

Keywords

- agroecology
- biodynamic agriculture
- ecoagriculture
- Evergreen Revolution
- farmer-led research
- food culture
- Green Revolution
- organic agriculture
- organic consumption
- organic farming
- organic transition
- small-scale producers
- sustainability
- sustainable agriculture
- sustainable food choice

Authors' Notes

Chapter 5
Acknowledgments
Funding in support of the oral presentation and publication of this paper was provided by OECD and USDA-NIFA. Conference assistance from the American Society of Agronomy–Organic Management Systems and ICROFS is appreciated. Core funding for the Iowa LTAR experiment through the Leopold Center for Sustainable Agriculture is gratefully acknowledged.

Chapter 6
Acknowledgments
Research on farmers' experiments in Austria and Cuba was funded by the Austrian Science Fund (FWF) under the grant number P19133 G14_01. Inputs for the paper came from the debate at the conference 'Innovations in Organic Food Systems for Sustainable Production and Enhanced Ecosystem Services' in Long Beach/California (USA) in November 2014. The participation of the first author at this conference was funded by the OECD's Co-operative Research Programme (CRP) on Biological Resource Management for Sustainable Agricultural Systems. We are also very grateful for the debate on our topic with Henrik Moller and for his suggestions to our manuscript.

Chapter 7
Acknowledgments
Funding from the European Community's Seventh Framework Programme (FP7/ 2007–2013) was received for the research leading to these results, under grant agreement no. FP7-266367 (SOLID-Sustainable Organic and Low Input Dairying). For further details, see www.solidairy.eu. We are also grateful for the active participation and support of all the farmers, researchers and SME partners.

Chapter 8
Acknowledgments
We would like to thank the Research Management Center (RMC) in UPM, which supported this study.

Chapter 9

Acknowledgments

The Sustainability Dashboard project is primarily funded by New Zealand's Ministry of Business, Innovation & Employment (contract AGRB1201), with additional co-funding from BioGro New Zealand, Zespri and Kiwifruit packhouse and orchard management companies, New Zealand Wine, and Te Rūnanga o Ngāi Tahu. We thank the OECD for funding to present our research findings at the Organic Food Systems for Sustainable Production and Enhanced Ecosystem Services symposium. We thank two anonymous reviewers for help suggestions on how to improve the manuscript.

Chapter 10

AFSIC staff member Becky Thompson provided invaluable technical assistance and support for this work. Special appreciation goes to Bill Thomas, current AFSIC Coordinator, Stephanie Ritchie, AFSIC Librarian, and former AFSIC coordinators, Jane Potter Gates and Jayne MacLean. Thanks also go to Andy Clark, Sustainable Agriculture Network coordinator, for his helpful review.

Index

A

Actinomycetes, 216, 240
Adoption
  process, 178
  rate, 62, 66
Agrecology movement, 118
Agricultural
  agencies, 142, 143
  approach, 136
  commodities, 250
  development, 59, 172
  disaster, 71
  ecology, 256
  economist, 252, 272, 279
  ecosystems, 265
  experiment station, 203, 213
  industry, 14
  innovation systems, 100, 131
  intensification, 53
  knowledge, 100, 112, 131, 132, 183
  landscapes, 55, 174
  methods, 261
  policy, 131, 174, 250, 276
  pollution, 262
  practices, 115, 137, 143, 203, 207, 219, 278
  practitioners, 294
  production, 55, 171, 176, 195, 247, 264, 265, 284
    intensification, 55, 171
    systems, 171, 176
  productivity, 53
    sustainability, 116, 118, 131
  professionals, 90, 135
  publications, 46
  research service (ARS), 80, 88, 94, 263, 282
  research, 109, 111, 116, 225, 289
  resource, 173, 248
  revolution, 195
  rural resource management, 172
  science/philosophy, 100, 275
  scientists, 69, 112, 274
  sector, 137, 141, 166
  soils, 93
  supply chains, 171
  system, 136, 151, 165, 190, 244, 256, 296
  technologies, 117
  testament, 219
Agriculture
  course, 34, 46, 49, 51
  dramatic mechanization, 9
  educators, 206
  food systems, 25, 94–96, 111, 113, 145, 224
  National Organic Standards Board, 272
  practices, 137
  research/education, 267, 272, 285
Agriecosystems, 154, 156
Agri-environmental policy, 262
Agro
  chemicals, 55, 172, 174, 291, 297
  ecological, 55, 73, 101, 118, 135, 174, 176, 180, 181
    concepts, 243
    location/management practices, 55
    paradigm, 174
    partnerships, 291
    principles, 89
  ecology, 102–104, 133, 169, 212, 256, 261, 262, 275, 290–294, 299
    movement, 102, 104
  ecosystem, 6, 54, 80, 89, 96, 101, 113, 261, 279, 290
    sustainability, 275
Agronomists, 109
Agronomy, 55, 96, 257, 297
Agro-tourism, 124
Alchemy/Threshold Generic Design Conference, 265

Alternative Farming Systems Information Center (AFSIC), 188, 245, 248, 255–259, 263, 267, 271, 272, 289, 292
American Society of Agronomy (ASA), 264
Anaerobic fermentation (pyrolysis), 72
Animal
derived proteins, 3
ecology/evolution, 211
fodder, 6
management, 123, 124
welfare, 14, 127, 148, 162
labour rights, 161
Anthropology, 10, 295
Archaeologists, 188
ARGOS
farms, 151, 152, 164
project, 148–154, 158, 164, 165
Artificial
circumstances, 18
preservatives, 17
Australian Organic Farming and Gardening Society (AOFGS), 32–39, 44–47
Avena sativa L., 86

B

Beltsville Agricultural Research Center (BARC), 88
Bengal rice famine, 71
Betteshanger Conference, 33
Binary comparisons, 156, 158, 166
Biodiversity, 6, 53–55, 60, 67, 71–76, 89, 122, 126, 128, 130, 135, 147–155, 161, 167–175, 180, 297
pest management, 261
Biodynamic
agriculture, 32, 34, 49, 50, 211, 299
farm, 249
farming/gardening, 32, 33, 45, 46, 50, 217
preparations, 217
Biodynamics Conference, 33
Bio-energy demand, 53
Bio-fertilizers, 74
Biogeochemical cycles, 290
Biological
activity, 54, 176
agriculture, 223, 225
balance, 70
control, 200, 203, 241, 242
ecological principles, 227
husbandry, 260, 261
organic unit, 217
output, 53
process, 55, 217, 238
sickness, 40
software industries, 76
Biomass, 64, 72, 84, 87, 90, 92, 95, 124, 153, 178
Biosecurity, 148
Biovillage paradigm, 72
Bromus inermis L, 86

C

California Certified Organic Farmers (CCOF), 83, 270
Cameron Highland, 139–143
Cameron Organic Produce (COP), 139
Cation exchange capacity (CEC), 91
Certification process, 142–144
CGIAR, 56, 112, 131, 267
Cheese-making farm, 125
Chemical
agriculture, 32
applications, 148, 153
farming, 33, 42
fertilizer, 55, 72, 135, 174, 195, 222
inputs, 72, 166
nutrient, 82
pesticides, 72
properties, 94, 196
reactions, 195
residues, 17
Chinese food philosophies, 13
Chisel-till (CT), 88, 263, 287
Civilizations, 183, 188, 230, 234
Commercial/industrial tendencies, 40
Common Objectives and Requirements of Organic Standards (COROS), 160, 162
Community/rural development, 289
Compost Manufacturers Manual, 217
Computer Models/Social Decisions, 246
Conservation, 56, 62, 70, 75, 76, 115, 140–143, 146, 152–155, 188, 190, 192, 197–203, 213, 227, 247, 251, 260, 266, 268

agriculture (CA), 27, 54–68, 111, 146, 176–182, 203, 214, 233, 247, 253, 270, 282
Agriculture Systems Alliance (CASA), 60
farming practices/flood control, 212
programs, 269
Controlling pests/disease, 296
Conventional
agriculture, 6, 22, 81, 96, 132, 254, 269
consumers, 22
dairy farms, 150, 152, 155
farmers, 158
farming/marketing, 225
systems, 81, 94, 95, 149, 159
food chain, 22
management, 80, 84, 91
methods, 254
rotation, 87, 91
system, 82–86, 88
wisdom, 189
Conventionalization, 6, 26
Cornucopia, 245, 259, 260
Cottony cushion scale, 199, 200
Creative Commons Attribution License, 31, 69
Critical race theory, 293
Crop
diversification, 64, 174
health, 89
management, 87, 176, 181, 202
nutrition/economics, 64
plants, 75
production, 27, 54, 56, 67, 83, 103, 174, 176, 182, 227
environment, 218
intensification, 54, 176
productivity, 76, 79, 90, 173
residues, 54, 176
rotations, 60, 81–87, 96
Science Society of America (CSSA), 264
seeds, 54
Cropping systems (CS), 85
C-sequestration, 180
C-stabilizing components, 89
Cuban
Agroecology Movement, 103
Urban Agriculture Movement, 103
Cultivating food justice, 293
Cultivation, 42, 54, 67–72, 75, 76, 86, 123, 221, 229
Cultivation/varieties of oats, 215
Cultural
context, 4, 24
dimensions, 5, 7
ecosystem services, 148
methods, 41
norm, 19
operations, 43
patterns, 4
recreation, 180
services, 148, 152, 153, 157
values, 4, 26, 128
Cyclopedia, 209

D

Dactylis glomerata, 86
Degradation, 53, 61, 136, 162, 172, 173, 176–179, 234, 295
Desertification, 189, 215
Diseases, 70, 71, 84, 173, 180, 239
Dutch, 5

E

Early American Soil Conservationists, 193, 195
Earth's Face and Human Destiny, 217
Earth's Green Carpet, 228
Earthworm density, 153
Ecoagriculture, 72, 75, 299
Eco-fisheries/eco-aquaculture, 72
Eco-friendly green label, 71
Eco-functional intensification, 156
Ecological
agricultural movement, 217
perspectives, 290
approach, 215
conscience, 232
constraints, 149
design, 266
effects, 257
energy subsidies, 151
general systems, 235

human resources, 188
processes, 72, 81, 153, 168
refuges, 155, 165
systems, 107, 148, 153–156, 160
Ecology
action, 247
biophysical features, 107
residue management, 180
Economic
ecological comparisons, 254
legal analysis, 262
plants, 263
progress, 252
resilience, 108
spiritual sicknesses, 40
Economics organic farming, 254, 276
Economics ethics ecology, 260
Ecosystem, 53, 54, 59, 89, 101, 128, 129, 147–159, 164–175, 180, 273
functioning, 158
services, 53, 54, 59, 89, 128, 129, 147–159, 164–167, 174, 175, 180
Ecotechnologies, 72–75
Ecoverification, 158, 166
Edmund Ruffin's efforts, 195
Effective Microorganisms (EM), 72
Emotional experiences/responses, 16
Empirical
evidence, 176, 179
experiences, 108
interview data, 5
studies, 106
Endangered Life-support Systems, 234
Endemic species, 154
Energy
costs, 65, 89
intensiveness, 248
recreation, 244
saving practices, 124
soil quality, 67
Environmental
accounting, 235
change, 28, 118
critics, 244
decision making, 235
degradation, 65
history, 107
movement, 225
performance, 159
problems, 269
production, 158
services, 54, 67, 89, 173, 177
social problems, 4, 23, 294
Environmentalists, 135
Eobiotic, 152
Erosion, 41, 43, 55, 61, 66, 67, 172, 175, 177, 180, 183, 192–198, 229, 239
Esoteric philosophy, 10
Ethical
principles, 9
production system, 159
Ethnobotany, 263
European Conservation Agriculture Federation (ECAF), 45, 46, 61
European Organic Production Statistics, 276
European Union (EU), 57, 99, 100, 110, 119, 132
Ever-green
agriculture, 76
revolution, 71–77, 299
Evolutionary process, 117
Experimentación campesina, 104
Experimental Circle of Anthroposophical Farmers and Gardeners (ECAFG), 45, 46
Experimentation, 104–106, 112, 113

F

Factor
analysis, 157
productivities, 53, 63, 175
productivity, 54
Factual position, 71
Fairness/futurity, 283
Fairness organic agriculture, 160
Family farming, 73
Family friendly farming, 286
Farm
diversification, 271
equipment, 56
garden digest, 35, 51
inspector trainings, 270
magazine, 271
management systems, 148

Farmer
engagement, 64
experiments, 100–111, 169
progress, 224
field schools (FFS), 65
led research, 126, 129, 299
Farming
gardening, 32, 35, 37, 46, 48, 226, 231
manual, 224
paradigms, 174
program, 126
system, 80–83, 94–96, 188, 255, 259, 266, 267, 271
approach, 164
Information Center, 188, 255, 259
trial (FST), 80–86, 93
Farmland Protection Policy Act, 255
Federal
agency, 213
entomological service, 199
Incentives and Farmers' Opinions, 277
Feeding practices/forage production, 123
Fertile soil, 219, 231
Fertility, 40, 43, 50, 57, 70, 82, 84, 89–93, 128–130, 210, 219, 224, 240, 297
farming, 238
gardening, 241
maintenance, 84
pastures, 238
Fertilization, 83, 89, 90, 221
Fertilizer, 69–72, 82, 116, 174, 175, 200, 202, 221, 294
Fertilizer application, 67, 239
Fibre production, 147, 149, 161, 165
Field Actions Science Reports, 53, 171
Financial
assistance, 140
capacity, 108
debt, 43
drivers, 108
indicators, 159
system, 40
Focus group discussion (FGD), 139
Food
Agricultural Policy, 260, 269, 285
Agriculture Conservation and Trade Act of 1990 (FACTA), 277, 291
alternativism, 10
choices, 4, 20, 23–26, 253
community regeneration, 281
consumption, 3, 7, 9, 16, 19, 21–26
culture, 296, 299
farming systems, 298
organization, 11
philosophy, 11
awareness, 14
connectedness with nature, 12
purity, 16
practices, 5, 6, 11, 16, 21–25
preparation, 14, 253
processing and distribution, 263
production, 23, 66, 144, 188, 189, 213, 221, 225, 244–247, 263, 266, 278
system, 144, 188
quality, 25, 89, 222, 248, 259, 263
security, 54, 71–76, 81, 115, 165, 168, 179–181, 252, 266, 267, 281, 289, 295–297
system, 4–9, 22, 110, 189, 253, 260, 279, 287, 293–297
Forest
degeneration and destruction, 226
problem, 226
resources, 266
Fragile ecosystems, 55
Fruit and vegetable growing, 240
Fruit orchards, 62
Fungi, 72, 95, 202, 240, 290
Furuno's system, 285

G

Game management, 232
Genetic diversity, 256
Genetic engineering, 261
Genetically modified (GM), 82, 86, 127
Genetics, 172
George washington carver, 205
German botanical association, 202
Germanic agricultural ideas, 33
Germplasm, 116, 119
Global
agricultural development, 252
area/regional distribution, 57

crop rotations/associations, 57
minimum soil disturbance, 57
organic soil cover, 57
citizen, 246
collapse, 246
communication system, 189
level, 180
perspective, 289
perspectives/developments, 171
warming, 75
Globalization of Agriculture and Food, 279
Globalized trade, 74
Grass productivity, 239
Grassland/animal production, 221
Grazing periods and forage production, 208
Green
agriculture, 72
manuring, 199, 212
revolution, 68, 70, 71, 75, 116, 174, 244, 299
Greenhouse
gas emissions, 54, 122, 180
gases (GHGs), 54, 67, 72, 73, 83, 122, 180
production, 274

H

Handbook of Medicinal Herbs, 262
Harvest of Doubt for the Garden State, 260
Health and Disease, 35, 48
Health and the New Civilization, 230
Health Land, 273
Health problems, 218, 295
Health organic agriculture, 159
Herbicides, 55, 57, 70, 72, 173, 174, 178
Herdsmanship, 238
High-external-input (HEI), 87
Holistic
management, 273
handbook, 273
production management, 6
research design, 108
resource management workbook, 273
Horticulture/plantation crops, 239
Human
activities, 235
degeneration, 214
diseases, 203
environment revolution, 237
health, 6, 48, 50, 225, 228, 232
nutrition, 203, 210, 228, 239, 243
resources development, 143

I

Indian Medical Service, 216
Indiana Food System, 260
Indica rice varieties, 70
Indigenous/scientific knowledge, 65
Indo Gangetic Plains, 63, 174
Industrial Agriculture, 284, 287
Industrialization, 8, 9, 278, 296
Industrialized nations, 298
Infiltration, 55, 60, 84, 175, 179
Information and communication technology (ICT), 72
Innovation systems, 112, 116, 119
Innovative
participatory approaches, 64
solutions, 119, 120
Inorganic
fertilizers/pesticides, 248
mineral fertilizers, 69
nitrogen fertilizer, 69
Integrated
management (IM), 148–155, 158–160, 165, 166
nutrient management (INM), 72
pest management (IPM), 72
Intermountain Research Station, 208
Internal Resources for Sustainable Agriculture, 289
International
Conference on Sustainable Agricultural Systems, 275
Ecosystem Health, 279
Federation of Organic Agriculture Movements (IFOAM), 4, 6, 27, 72, 132, 136, 146, 159–168, 223, 252, 261, 274
policy, 298
relations, 42
Society for Ecology and Culture (ISEC), 287
survey, 229

Iodine, 73, 125, 126
Irish potato famine, 71
Iron, 73
Irrigation, 70, 126, 198, 203, 207, 227

J

Japonica rice varieties, 70
Journalistic investigation, 296

K

Kiwifruit orchards, 149–155, 169
Knowledge
  construction, 129
  empowerment, 72–74
  information system, 100

L

Labour costs, 142
Labour in Agriculture, 229
Labour shortages, 141, 143
Land
  issue, 140, 143
  landscape, 107
  management, 280
  resource, 53, 173
  robbery, 221
  sharing, 152, 165
Large-scale monoculture, 43
Lazy-bed Gardening, 247
Learning and Innovation Networks for Sustainable Agriculture (LINSA), 129
Lebensreform, 8–10, 21, 26
Livestock and Human Nutrition, 218
Living Soil Association of Tasmania (LSAT), 47
Long Term Research on Agricultural Systems (LTRAS), 85
Low-external-input (LEI), 87
Low-input/Sustainable Agriculture (LISA), 272, 277

M

Maine Food System, 260
Malaysian
  agricultural sector, 141
  farmers, 136, 137
  Organic Certification Program, 138
  small-scale producers, 137
Manual for State Food Systems Analysis, 260
Marginal farming, 71, 74
Market
  conditions, 118
  flagship, 153
  pressures, 61
Material
  considerations, 42
  qualities, 18
Meat consumption, 14
Mechanical
  devices, 43
  management, 178
Medicinal/culinary herbs, 263
Meloidogyne javanica, 84
Meta-hypothesis, 155, 156
Methane, 72, 181
Metropolitan areas, 269, 283
Microactivities, 240
Microbial
  biomass, 84
  community, 84
Microorganisms, 72, 203, 240
Microsoft-Access database, 103
Milk production, 74, 131, 132, 150, 280
Millenium Ecosystem Assessment, 157
Millet/sorghum, 123
Mineral
  fertilizer, 175
  nutrition, 238
  organic origin, 55
Ministry of Business Innovation and Employment (MBIE), 164
Modern information/communication, 72–74
Molecular biology, 297
Monocropping, 224
Mulch Gardening, 237
Mulching of Vegetables, 239
Multidimensional scaling, 158, 159
Multi-generational Home based Business Testament, 286
Multiple Cropping Systems, 267, 289
Mycorrhiza, 202

N

N mineralization, 84, 88, 90
Nano modern biotechnology, 73
Nanotechnology, 152, 163
National
  Academy of Sciences Board on Agriculture, 274
  Academy of Sciences Book, 272
  Academy Press, 272, 274, 280
  Agricultural Library, 187, 190, 197–200, 210–216, 218, 221, 224, 225, 229, 230, 255
  Agricultural Library Digital Repository (NALDR), 199, 210, 229
  agricultural policies, 137
  health, 216
  menace, 213
  organic farmers' survey, 282
  organic program (NOP), 224, 276, 277, 291
  policies/legislation, 124
Native American Agriculture and Wild Plant Conservation, 268
Natural
  appearance, 18
  biological processes, 55
  capitalism, 251
  circumstances, 14
  ecology, 221
  enemies, 200, 241, 242
  environment, 12, 13, 23, 174
  foods primer, 242
  health foods, 10
  lifestyle, 22
  medicine, 8
  phenomena, 7
  resource, 54, 64, 123, 127, 130, 182, 188, 260
Natural resource
  base/environment, 54
  conservation, 289
  management, 72, 123
  resource, 53, 72–74, 115, 119, 171, 172, 182, 193
    management systems, 72
    conservation service (NRCS), 213
  sources, 119, 123
  way of farming, 267
New Jersey Food System, 260
New York State Food System, 260
New Zealand
  agriculture, 164, 169
  exported products, 164
  organic growers, 165
N-fertilizer, 64
Nitrogen
  fertilizers, 152
  fixation and soil fertility, 210
Non-governmental
  agencies, 74
  programs, 255
Nonorganic counterparts, 153
North Central Regional Conference, 267, 289
Northbourne mentions, 37, 38, 44
No-till (NT), 37, 88
Nutrient
  cycles, 180
  efficiency, 67
  water supply, 290
Nutrition
  debate, 253
  health, 215
  security, 73, 75
Nutritional
  needs, 41
  outcomes, 73
  studies, 218
  value, 124

O

Okada's farming system, 230
Oklahoma's future, 285
Omnivore's dilemma, 290
Operational groups, 118
Orchardgrass, 86
Organic
  agricultural movement, 290
  agriculture, 6, 25–31, 34, 37, 49, 50, 72, 82, 95, 101, 127–130, 135–138, 141–146, 152–158, 167–169, 195, 229, 243, 247, 259, 274, 299
  Centre Wales, 276

certification, 80, 90, 93, 150, 165
conditions, 123
consumers, 4–7, 10–12, 20, 22–25, 83
consumption, 4, 10, 24, 299
conventional crops, 90
crop, 85
rotations, 87
dairy farming, 124
farms, 155
environmental movement, 10
Farm Management Handbook, 276
farmers, 26, 80, 83, 89, 101–104, 108–111, 116, 118, 129, 130, 136–144, 157, 158, 162, 166, 169
Farming, 32–39, 44–48, 50
research foundation (OFRF), 282
transitions, 224
farms, 14, 116, 120, 138, 142, 150, 152, 155, 156, 161, 165, 273, 296
food, 4–11, 20, 22–29, 132, 138, 295, 298
consumption, 6, 11, 23, 25, 28
Philosophy, 3, 5–13, 15–21, 23–29
Foods Production Act (OFPA), 276, 277
Gardening, 226, 231, 233, 237, 241, 245
ingredients, 23
inputs (OI), 87
management, 81, 91, 167, 169
matter, 43, 64, 86, 87, 90, 92, 96, 125, 172, 175, 176, 197, 214
meat, 14, 16, 19
method, 150, 155, 226, 231, 235, 276, 289
movement, 5–10, 20, 25, 109, 149, 158, 166, 226
natural foods movement, 7
philosophy, 21, 24, 222
plots, 82, 83, 90
practices, 80–84, 88, 143, 188
premium price, 83
price premiums, 83, 87
production, 6, 25, 89, 90–93, 137, 138, 143, 152, 163, 166, 220, 241, 297
products, 5, 137, 139–142
research and development institutions/agencies, 144
resource network, 280
rotations, 88–91
scheme, 138, 145
standards, 24, 130, 138, 148, 149, 150, 155, 159–166
system, 81–91, 124, 151, 153, 167, 282
transition, 93, 299
vegetables, 274
versus chemical farming, 32, 40

P

Packing/storage facilities, 142
Participant/sampling procedures, 139
Pedagogic method, 73
Pedo-climatic conditions, 128
Pennsylvania Food System, 260
Pest
control, 175, 253, 257, 262, 297
management, 54, 81, 82, 90, 149, 169, 174, 176, 289
problems, 175
suppression, 89
Pesticides, 55, 70–74, 82, 90, 101, 135, 151, 155, 168, 173–175, 180, 181, 241, 244, 268, 278, 285, 294
Pfeiffer mentions, 37, 38
Phleum pratense L, 86
Pisum sativum L, 86
Plant
development, 240
nutrient management, 54
nutrition, 174, 195, 202, 240
plant interactions, 193
pollinating organisms, 268
respiration, 194
Plundered planet, 230
Polarization, 22, 224, 225
Policies/institutional support, 59
Policy
analysis, 262
framework, 102
institutional support, 64, 66, 178
makers, 24, 56, 135, 137, 144, 164, 166, 178–181, 189, 258, 284
Regulatory Environment for Organic Farming, 276
research activity, 258

Political
    conditions, 115
    context, 230
Pollination, 180
Pollution-mitigation methods, 89
Poor quantification, 107
Post-war reconstruction, 222
Pottenger's Cats, 214
Poultry manure, 89
Private/civil sector support, 66
Product differentiation/marketing, 123–125
Production
    costs, 142, 175
    intensification, 172, 174
    performance, 159
Productivity
    profitability, 64
    sustainability, 74
Promotion versus prevention, 5
Protein rich crops, 123
Pro-women/proemployment orientation, 73

Q

Q sort methodology, 127
Qualitative
    content analysis, 102
    data, 139
    exploration, 3
    interview, 5, 29
Quasi-experimental evidence, 150
Queensland Country Life, 37, 39, 50

R

Range/pasture management, 208
Rebuild forest resources, 226
Recycling materials, 266
Reform movements, 8, 22, 23
Regenerating agriculture, 111, 281
Regenerative
    agriculture association, 270, 271
    food system, 267, 272
Research
    designs, 108
    process, 105, 108, 109
Rethinking management, 265
Revere clay loam, 86
Revolution, 56, 68–77, 135, 174, 224, 230, 244, 251–253, 268, 285, 286
Rice production, 63, 181
Robustness, 107
Rodale Institute (RI), 81–83
Rodolia cardinalis, 200
Romantic
    religions, 8
    worldview, 21
Roots of Productive Conservation, 260
Rotation/association, 57
Rothamsted experiments, 206
Rural resource management, 181

S

Salinity problems, 67
Science/engineering, 254
Science/Environmental Health Network, 258
Scientific
    achievements, 240
    approach to organic farming, 260
    geographical discoveries, 193
    institutions, 74
    knowledge, 129, 193
    natural history, 211
    practices, 294
    symposium, 264
Self-help groups (SHGs), 74
Semi-arid Mediterranean environments, 64
Semi-structured interviews, 102, 103, 157
Silent spring, 10, 241
Slash/mulch rotational farming, 176
Small
    farm development, 267
    scale grain raising, 245
    scale producers, 136, 138, 140, 143, 299
Social
    forces, 22
    learning, 129
    legislation, 40
    orientations, 158
    relationship, 142
    societal innovation, 117
Societal/political goals, 117
Socio-cultural/economic conditions, 101
Soft
    acidity, 90

aeration, 60
civilization, 234
energy paths, 251
health foundation, 203
library, 195, 200–206, 210–216, 219–225, 227–231, 234, 237–240, 249, 268
Soil
Association, 35, 48, 49, 126, 203, 219, 222, 223, 232, 247
Association of South Australia (SASA), 47, 49
biopores, 179
compaction, 172
conservation practices, 212
society, 260
cover, 54, 57, 64, 177
degradation, 173, 179
disturbance, 54, 57, 176
erosion, 40, 42, 44, 48, 55, 173, 179, 212, 213, 218, 229, 262, 297
fertility, 44, 49, 50, 206, 210, 217, 238, 243, 249, 284
animal husbandry, 238
renewal/preservation, 50, 217
formation, 41, 173, 214
function, 91
health, 48, 71, 89, 91, 152, 172–178
microbial abundance, 84
microbiologist, 243
microbiology, 216
microorganisms and higher plants, 240
moisture, 179
nutrients, 84
organic carbon (SOC), 88–92
organic matter (SOM), 64, 81, 123, 125, 138, 142, 145, 172, 175, 214
properties, 214
protection, 56, 67
quality, 64, 79–84, 88–94, 123
resource, 214
Science Society of America (SSSA), 264
structure, 70, 172–176
system, 173
tillage, 56, 175
type, 80–85, 90, 94
vitality, 223
water, 60, 84
plant, 227
Solar ecology, 265
Soya crops, 61
Soya monocropping, 61
Spatial scale (synchronic strength), 107
Split-plot design, 84
Starvation, 40, 297
State Library of South Australia (SLSA), 47, 49
Steiner's injunction, 34
Strip tillage, 57
Sustainability Assessment of Food & Agriculture (SAFA), 148, 149, 160–165, 168
Sustainable Agriculture Research and Education (SARE), 272, 277, 279, 291
Sustainable Food Systems, 263, 275, 299
Sustainable Organic Low-Input Dairying (SOLID), 115–119, 123–129
Sustainable production intensification, 54, 67, 176, 180, 182
Symbiotic nitrogen fixation, 290
Synchronic weakness, 107
System of Rice Intensification (SRI), 176, 181, 183

T

Taylor's framework, 24
Technical
capacity, 178
optimism, 117
possibilities, 66
scope, 291
Technological
issues, 288
optimism, 116
Temporal replication, 107
Temporary Occupation Licenses (TOLs), 140
Theoretical concepts, 4, 56
Thermic Typic Xerothents, 83
Topography, 149, 214
Tourism activities, 136
Toxic residues, 70
Toxicology, 297

Traditional agriculture, 71
Transgenic organisms, 148, 163
Tree crops, 213
Trees/toadstools, 225

U

U.S. Department of Agriculture (USDA), 80–82, 88, 94–96, 207, 209, 212, 213, 224, 242, 250, 255, 263, 272–279, 282, 283, 285
Unifying concepts, 265
United Nations Food and Agriculture Organisation, 148
United States Agency for International Development (USAID), 275
Universal perspective, 268
Unwelcome harvest, 281
Urban
  agriculture, 281
  development, 23, 251, 255
  environment, 281
  food production issues, 263
  future, 266
Urbanization, 8, 9

V

Vanishing lands, 218
Variable Input Crop Management Systems (VICMS), 80, 86, 87, 93
Vedalia cardinalis, 200
Vegetable waste, 219
Vegetarian
  diets, 8
  food, 14
  meals, 14
Vegetarianism, 9
Vegetative associations, 208
Verbatim, 12
Vermiculture, 74
Vetiver grass, 244
Village knowledge centres (VKCs), 72, 74
Vineyards/olive plantations, 62
Violent disruption, 227
Vitamin A, 73
Vitamin B, 73
Voisin Grazing Management, 270

W

War Economy Standard, 45, 49
Water
  agriculture, 265
  availability/consumption, 265
  erosion, 177
  every farm, 237
  management, 68, 183, 260, 265
  management/policy, 262
  pollution, 135, 297
  quality, 55, 81, 82, 89, 93, 167
  tables, 252
Wealthy farmers, 143
Weed
  control, 67, 93, 174, 220
  management, 64, 86, 87, 91, 93
  water management, 54, 55
Wheat/rice, 70, 71
Wheat/maize cropping, 57
Wheat varieties, 285
White agriculture, 72
Winrock International, 263, 264
Winter cover crops, 84
Woody vegetation, 155
Working draft, 260
World agriculture, 227
World survey, 218
World War II, 31, 220–222, 228, 230
Worldwatch Institute Report on Progress, 266

X

Xenobiocides, 152
Xenobiotic, 152

Y

Yearbook of Agriculture, 217
Yolo silt loams, 83

Z

Zambia/Zimbabwe, 66
Zero-external-input (ZEI), 87
Zinc, 73